Experimental Approaches for the
Investigation of Innate Immunity
The Human Innate Immunity Handbook

Experimental Approaches for the
Investigation of Innate Immunity
The Human Innate Immunity Handbook

Editors

Ruth R Montgomery
Richard Bucala

Yale University School of Medicine, USA

We World Scientific

NEW JERSEY • LONDON • SINGAPORE • BEIJING • SHANGHAI • HONG KONG • TAIPEI • CHENNAI • TOKYO

Published by

World Scientific Publishing Co. Pte. Ltd.
5 Toh Tuck Link, Singapore 596224
USA office: 27 Warren Street, Suite 401-402, Hackensack, NJ 07601
UK office: 57 Shelton Street, Covent Garden, London WC2H 9HE

Library of Congress Cataloging-in-Publication Data
Experimental approaches for the investigation of innate immunity : the human innate immunity
handbook / edited by Ruth R. Montgomery, Richard Bucala.
  p. ; cm.
  Includes bibliographical references and index.
  ISBN 978-9814678728 (hardcover : alk. paper)
  I. Montgomery, Ruth R. (Ruth Rebecca), editor. II. Bucala, Richard, editor.
  [DNLM: 1. Immunity, Innate. 2. Autoimmunity. 3. Communicable Diseases--immunology.
4. Immunologic Techniques--methods. QW 541]
  RC582
  616.07'9--dc23

                                                                                2015033002

British Library Cataloguing-in-Publication Data
A catalogue record for this book is available from the British Library.

Cover:
Healthy human blood cells were arranged by the viSNE algorithm based on per-cell expression of
27 proteins measured by mass cytometry.
Illustration by Jonathan M. Irish (Vanderbilt University School of Medicine) with thanks to
Christopher O. Ciccolella (Cytobank Inc.).

Typeset by Stallion Press
Email: enquiries@stallionpress.com

Contents

Preface

Experimental Approaches for Investigation of Innate Immunity: *The Human Innate Immunity Handbook.*

World Scientific Press, 2016

Human populations and individuals have varied immune status and function that contribute to the diversity of disease severity and responses to infection. Variability in immune responses may be due to genetic polymorphisms in key recognition and response proteins, and the acquired influence of historical environmental exposures in the host. Such variation has long been appreciated to be an inherent feature of adaptive immune responses, where HLA-restricted pathways may predominate, but it has become increasingly clear that there is important and clinically meaningful variation in innate responses as well. The recent explosion of information in innate immune pathways for cellular recognition, effector responses, and their genetic regulation has prompted new investigations into analogous pathways in the human immune response. Perhaps the greatest challenge facing the investigation of the human immune response is the inter-individual variation in measured responses and the integration of collected data into a unified and unambiguous set of results. Building a deeper understanding of human immune function relies on focused methods with rigorous controls and quantitative analysis.

The Human Innate Immunity Handbook gives readers up-to-date information across a spectrum of experimental approaches for analyzing such complex responses at molecular, cellular, and multicellular levels, and includes studies of both model and *in vivo* systems. This volume, edited by two practitioners in the field of human immunology, provides a compendium of up-to-date methodologies for the investigation of human innate immunity with application to the study of normal immune function, autoimmunity, infectious diseases, and immunosenescence. Microfluidic and high throughput approaches, advanced cytometry, and bioinformatics are covered. Chapters include quantitative flow cytometry for Toll-like receptor expression and function; multidimensional single cell mass cytometry, Imagestream high resolution microscopy coupled to flow cytometry, the study of cell interactions in biomimetic organs, high-throughput single cell secretion profiling, multiplexed transcriptomic profiling, and RNA interference and microRNA methodologies. The insights gained offer new understanding into both normal physiology and immunopathology. We hope that this volume can provide a useful single resource for investigators that may assist in the continued advancement of knowledge in human innate immunity.

Ruth R. Montgomery
Richard Bucala

Yale University School of Medicine
New Haven, CT

List of Contributors

Heather Allore, PhD
Yale University School of Medicine
Department of Internal Medicine
Section of Geriatrics
New Haven, CT USA
Email: heather.allore@yale.edu

Catherine Blish, MD, PhD
Stanford University School of Medicine
Department of Medicine
Division of Infectious Diseases and Geographic Medicine
Stanford, CA USA
Email: cblish@stanford.edu

Chris Cotsapas, PhD
Yale University School of Medicine
Department of Neurology and Genetics
New Haven, CT USA
Email: chris.cotsapas@yale.edu

Rong Fan, PhD
Yale University
Department of Biomedical Engineering
New Haven, CT USA
Email: rong.fan@yale.edu

Allison R. Greenplate, BS
Vanderbilt University School of Medicine
Department of Pathology, Microbiology and Immunology
Department of Cancer Biology and Vanderbilt-Ingram
Cancer Center
Nashville, TN USA
Email: allison.r.greenplate@vanderbilt.edu

Jose D. Herazo-Maya, MD
Yale University School of Medicine
Department of Pulmonary, Critical Care and Sleep Medicine
New Haven, CT USA
Email: jose.herazo-maya@yale.edu

Erica Herzog, MD PhD
Yale University School of Medicine
Department of Internal Medicine
Section of Pulmonary, Critical Care and Sleep Medicine
New Haven, CT USA
Email: erica.herzog@yale.edu

William Housley, PhD
Yale University School of Medicine
Department of Neurology
New Haven, CT USA
Email: william.housley@yale.edu

Daehee Hwang, DGIS
Institute for Basic Science
Department of New Biology & Center for Plant Aging Research
DGIST, Daegu, Korea
Email: dhhwang@postech.ac.kr

Jonathan M. Irish, PhD
Vanderbilt University School of Medicine
Department of Cancer Biology and Vanderbilt-Ingram Cancer
Center

Department of Pathology, Microbiology and Immunology
Nashville, TN USA
Email: jonathan.irish@vanderbilt.edu

Wan-Uk Kim, MD PhD
The Catholic University of Korea
Department of Internal Medicine
Division of Rheumatology
Seoul, South Korea
Email: wan725@catholic.ac.kr

Ruth R. Montgomery, PhD
Yale University School of Medicine
Department of Internal Medicine
Section of Rheumatology
New Haven, CT USA
Email: ruth.montgomery@yale.edu

Ewa Menet, MS
Yale University School of Medicine
Department of Laboratory Medicine
New Haven, CT USA
Email: ewa.menet@yale.edu

Katherine Miller-Jensen, PhD
Yale University
Department of Biomedical Engineering & Molecular,
Cellular & Developmental Biology
New Haven, CT USA
Email: kathryn.miller-jensen@yale.edu

Subhasis Mohanty, PhD
Yale University School of Medicine
Department of Internal Medicine
Section of Infectious Diseases
New Haven, CT USA
Email: subhasis.mohanty@yale.edu

Feng Qian, PhD
Fudan University
State Key Laboratory of Genetic Engineering and Ministry
of Education Key Laboratory of Contemporary Anthropology
School of Life Sciences
Shanghai, China
Email: fengqian@fudan.edu.cn

Mikael Roussel, MD, PhD
Hematology Laboratory, University Hospital, Rennes, France
Department of Pathology, Microbiology & Immunology & Cancer
Biology
Department of Cancer Biology and Vanderbilt-Ingram
Cancer Center
Vanderbilt University School of Medicine
Nashville, TN USA
Email: mikael.roussel@chu-rennes.fr

Albert C. Shaw, MD, PhD
Yale University School of Medicine
Department of Internal Medicine
Section of Infectious Diseases
New Haven, CT USA
Email: albert.shaw@yale.edu

Arpita Singh, PhD
Yale University School of Medicine
Department of Neurology and Genetics
New Haven, CT USA
Email: arpita.singh@yale.edu

Dara M. Strauss-Albee, BA
Stanford University School of Medicine
Department of Immunology and Medicine
Stanford, CA USA
Email: darasa@stanford.edu

Huanxing Sun, PhD
Yale University School of Medicine
Department of Internal Medicine
Section of Pulmonary, Critical Care and Sleep Medicine
New Haven, CT USA
Email: huanxing.sun@yale.edu

Jose Thekkiniath, PhD
Yale University School of Medicine
Department of Internal Medicine
Section of Rheumatology
New Haven, CT USA
Email: jose.thekkiniath@yale.edu

Mark Trentalange, MD, MPH
Yale University School of Medicine
Department of Internal Medicine
Section of Geriatrics
New Haven, CT USA
Email: mark.trentalage@yale.edu

Adrian Wyllie, MD
Yale University School of Medicine
Department of Internal Medicine
Section of Pulmonary, Critical Care and Sleep Medicine
New Haven, CT USA
Email: adrian.wyllie@yale.edu

Yi Yao, PhD
Yale University School of Medicine
Department of Internal Medicine
Section of Rheumatology
New Haven, CT USA
Email: yi.yao@yale.edu

Sungyong You
Samuel Oschin Comprehensive Cancer Institute
Cedars-Sinai Medical Center
Departments of Surgery & Biomedical Sciences
Division of Cancer Biology and Theurapeutics
Los Angeles, CA USA
Email: sungyong.you@cshs.org

Yangyang Zhu, BS
Yale University School of Medicine
Department of Internal Medicine
Section of Pulmonary, Critical Care and Sleep Medicine
New Haven, CT USA
Email: yangyang.zhu@yale.edu

Chapter 1

Assessment of Toll-Like Receptor Expression and Function by Flow Cytometry

Subhasis Mohanty[*,†] *and Albert C. Shaw*[*,‡]

*Yale School of Medicine, Section of Infectious Diseases,
300 Cedar St., P. O. Box 208022, New Haven, CT 06520 USA*
[†]*subhasis.mohanty@yale.edu*
[‡]*albert.shaw@yale.edu*

Toll-like Receptors (TLRs) are germline encoded, membrane-associated pattern recognition receptors (PRRs) of the innate immune system that recognize pathogen-associated molecular patterns and couple the innate and adaptive immune responses.[1-3] Almost two decades after Janeway put forth the hypothesis of "pattern recognition" and a decade and half after their discovery as PRRs, TLRs have been shown to execute a pivotal role in immune functions of multicellular organisms.[4-6] Assessments of TLR function generally requires treatment of cells with defined TLR ligands — such as triacylated lipoproteins recognizing TLR1/2, diacylated lipoproteins recognizing TLR2/6, double-stranded (ds) RNA (TLR3), lipopolysaccharide (TLR4), flagellin (TLR5), single-stranded (ss) RNA (TLR7 or TLR8) and unmethylated CpG-rich DNA (TLR9). Here, we describe considerations for

---

[†]Corresponding author.

evaluating human TLR function in peripheral blood mononuclear cells (PBMCs) using intracellular cytokine staining to evaluate TLR-induced cytokine production in specific cell lineages.[6,7]

1. Sample Preparation

Cryopreserved samples of peripheral blood or isolated peripheral blood mononuclear cells (PBMCs) are frequently employed in human immunology studies. The use of such samples has obvious advantages, allowing investigators to analyze large numbers of samples from well-characterized cohorts. In addition, several studies indicate that analyses of *B* or *T* cell populations appear to be reasonably stable following freeze-thawing.[8,9] However, for other cell types such as monocytes, dendritic cells (DCs), NK cells and especially neutrophils, blood samples should ideally be processed immediately to obtain best results. For example, Toll Like receptor (TLR)-induced cytokine production in monocytes or DCs is affected by both delay in processing and cryopreservation.[10] Consequently, we try to standardize all processing procedures, at a minimum providing a maximal window of time for processing and ideally keeping unavoidable delays in sample processing uniform for all subjects.

1.1 *Sample collection*

- Collect 4 or more 10 mL tubes of blood (BD Vacutainer Sodium Heparin tubes) per subject (expected yield of PBMCs is approximately $1.5–2.0 \times 10^6$ per mL of blood).
- Transfer the blood tubes to a tissue culture hood.
- Dilute blood with equal volume of 1× Sterile PBS (Dulbecco's Phosphate Buffered Saline 1×).

1.2 *Gradient centrifugation for isolation of PBMCs*

- Layer diluted blood over 15 mL Histopaque (SIGMA 1077) in a 50 mL Polypropylene Conical tube. (Thus 2 × 50 mL tubes for centrifugation per individual)
  (Note: Histopaque should be stored at 4°C. It can be filtered through a 0.22 μm Cellulose Acetate filtration unit and 15 mL

aliquots stored at 4°C in 50 mL polypropylene conical tubes. However, prior to layering with diluted blood Histopaque must be thawed to room temperature).

- Centrifuge for 20–30 min at 400 × g in a swinging bucket centrifuge without brake at 22°C.

- Carefully transfer the tubes to the tissue culture hood. The PBMCs are in a white ring, the buffy coat, at the interface between Histopaque and medium. Aspirate top layer and stop a few mL above the buffy coat.

- Collect the buffy coat by use of a sterile polyethylene transfer pipette to another 50 mL Polypropylene Conical tube containing 20 mL of cold 1× HBSS (Hanks' Balanced Salt Solution). Place the tubes on ice/cold until further use.
  (Note: HBSS aliquots of 20 mL each can be prepared ahead of time and kept on ice until use).

- Adjust the volume of each tube containing PBMCs to 50 mL with cold 1× HBSS. Centrifuge the tubes at 300 × g at 4°C for 10 min to pellet PBMCs.

- Remove the supernatant and re-suspend the pellets in 50 mL of cold 1× HBSS and place on ice.

- Take a 25 µL aliquot and dilute with 25 µL of trypan blue and obtain cell count using a hemocytometer. Alternatively, cells can be counted using cell counter instruments or a flow cytometer such as the Guava EasyCyte using the ViaCount kit.

- Centrifuge cells at 300 × g at 4°C for 10 m to pellet PBMCs and re-suspend in complete RPMI 1640 medium (with 10% Fetal Bovine Serum and 10U/mL Penicillin/Streptomycin) at $1 \times 10^7/$ mL and store in ice.

- PBMCs can now be used for downstream applications. For assays of innate immune function, we use freshly isolated PBMCs only.

2. Assessment of Human TLR Function

2.1 *TLR Expression on monocytes and DCs*

Monocytes express surface TLR1, TLR2, TLR4 and TLR5, and TLR 3 and TLR 8 in endosomes. Similarly, DCs also express TLR1, TLR2, TLR4 and TLR5 on the cell surface and endosomal TLR 3, TLR7/8 and TLR 9. Consequently, endosomal TLRs require assessment via intracellular staining following surface staining for lineage markers. Please refer to table 1 for complete list of antibodies. Each lot of antibodies should be titrated to determine an optimum working concentration.

2.1.1 *Surface Staining for TLR Expression*

- Dispense 100 μL of the freshly isolated PBMC suspension (a minimum of 1×10^6 PBMCs to analyze DC subsets or at least 0.5×10^6 PBMCs for analysis of monocyte subsets) in each of two wells of a 96-well round bottom tissue culture plate (with one well used for isotype control) per subject.
- Centrifuge at $300 \times g$ for 10 min at 4°C.
- Discard the supernatant by flicking plates upside down and place plates on ice.
- Add 200 μL of ice-cold FACS buffer ($1 \times$ PBS containing 0.1% FBS) to each well and re-suspend the pellet.
- Centrifuge cells at 300 x g for 10 min at 4°C.
- Add 100 μL of antibody cocktail (at $2 \times$ final concentration) to respective wells and mix by pipetting using a multichannel pipettor, and incubate plate on ice or 4°C for 20–30 min (For DCs, use Table 1, panels C and D, excluding antibodies recognizing the endosomal TLRs 3, 7, 8 and 9).
- Add 100 μL of $1 \times$ FACS buffer and centrifuge at $300 \times g$ for 10 min at 4°C.
- Remove the unbound antibody by discarding the supernatant by flicking as above.
- Re-suspend pellets in 100 μL per well of 3% paraformaldehyde in PBS.
- Incubate on ice or 4°C for 10–20 min.
- Centrifuge plates at $300 \times g$ for 10 min at 4°C.
- Discard excess paraformaldehyde.
- Re-suspend the cell pellet in 100μL of freezing medium (10% DMSO in FBS).
- Place the plates at −80°C until acquisition using the flow cytometer.

Table 1. Flow Cytometry Antibody Panel for Human Dendritic Cells.

Fluorochrome	FITC	PE	PE-Texas Red	PerCP	Alexafluor700	PE-Cy7	APC	APC-Cy7	Pacific Blue
A. Dendritic Cell Intracellular Cytokine Panel	Anti-IFNα (MMHB-3)	Anti-IL-12p70 (20C2)	Anti-CD3 (S4.1, AKA7D6) Anti-CD19 (SJ25-C1) Anti-CD14 (TuK4) Anti-CD16 (3G8)		Anti-IL-6 (MQ2-13A5)	Anti-CD11c (3.9)	Anti-CD123 (7G3)	Anti-HLA-DR (LN3)	Anti-TNFα (Mab11)
B. Dendritic Cell Co-Stimulatory Panel	Anti-CD80 (2D10.4)	Anti-CD86 (IT2.2)	Anti-CD3 (S4.1, AKA7D6) Anti-CD19 (SJ25-C1) Anti-CD14 (TuK4) Anti-CD16 (3G8)		Anti-CD40 (5C3)	Anti-CD11c (3.9)	Anti-CD123 (7G3)	Anti-HLA-DR (LN3)	Anti-iCOS-L (MIH12)

(Continued)

Table 1.   (*Continued*)

Fluorochrome	FITC	PE	PE-Texas Red	PerCP	Alexafluor700	PE-Cy7	APC	APC-Cy7	Pacific Blue
C. Dendritic Cell TLR Panel-I	Anti-TLR5 (85B152.5)	Anti-TLR3 (11F8)	Anti-hCD3 (S4.1, AKA7D6) Anti-CD19 (SJ25-C1) Anti-CD14 (TuK4) Anti-CD16 (3G8)	Anti-hTLR2 (383936)	Anti-hTLR8 (T2.5)	Anti-CD11c (3.9)	Anti-CD123 (7G3)	Anti-HLA-DR (LN3)	Anti-TLR1 (TR23)
D. Dendritic Cell TLR Panel-II	Anti-TLR8 (T2.5)	Anti-TLR9 (eB721665)	Anti-CD3 (S4.1, AKA7D6) Anti-CD19 (SJ25-C1) Anti-CD14 (TuK4) Anti-CD16 (3G8)	Anti-TLR7 (533707)	Anti-TLR4 (HTA125)	Anti-CD11c (3.9)	Anti-CD123 (7G3)	Anti-HLA-DR (LN3)	

2.1.2 *Intracellular staining for TLR expression*

- Process freshly isolated PBMCs for surface staining as in 2.1.1 above.
- On the day of analysis, remove sample plates from −80°C and thaw on ice.
- Add 100 μL of 1 × FACS buffer (1 × PBS with 0.1% FBS) to each well and mix
- Centrifuge at 300 × g for 10 min at 4°C and discard the supernatant by flicking the plate.
- Add 100 μL of CytoFix/CytoPerm solution (BD Biosciences).
- Incubate for 20 min in the dark at 4°C.
- Centrifuge plates at 300 × g for 10 min at 4°C and discard the supernatant.
- Add 100 μL of 2 × antibody cocktails for intracellular TLRs (TLR 3, TLR7, TLR8 and TLR9) in 1 × BD perm/wash buffer to each well.
- Incubate 20–30 minutes on ice or 4°C in the dark.
- Add 100 μL of 1 × BD Perm/Wash buffer to each well and centrifuge for 10 min at 4°C at 300 × g.
- Discard the supernatant.
- Re-suspend cell pellets in 200 μL of 1 × FACS buffer and analyze immediately.

3. Assessment of Toll-like Receptor Function

Synthetic ligands such as triacylated lipoprotein Pam3CysSerLys4 (Pam3CSK4) (10μg/mL) recognized by the TLR 1/2 heterodimer, lipoteichoic acid (LTA) (2μg/mL) for the TLR2/6 heterodimer, dsRNA or poly (I-C) (10μg/mL) for TLR3, lipopolysaccharide (LPS) (1μg/mL) for TLR4, Flagellin (5μg/mL) for TLR5, R848 (imiquimod) (0.5μg/mL) for TLR 7 and 8 and CpG DNA (3μg/mL) for TLR9 can be used in tissue culture at varying concentrations to stimulate specific TLRs. A list of TLR ligands and working concentrations used for PBMC stimulation can be found Table 3. Note for TLR9 stimulation that it is important to utilize CpG oligodeoxynucleotides of the class mediating type I interferon production (type A or C), as opposed to *B* cell stimulation (type B).[11] We have used TLR ligands from Invivogen for our experiments. Stimulation-induced changes in expression of co-stimulatory proteins such as CD80, CD86, CD40, or CD62L, or production of intracellular cytokines such as IL-6, TNFα, IFNα/β, IL-10, IL-12p40 or p70 can be estimated utilizing multidimensional flow cytometry.

3.1 *TLR ligand induced co-stimulatory molecule expression in monocytes and dendritic cells*

3.1.1 *In-vitro stimulation of monocytes or DCs for co-stimulatory molecule expression*

- Incubate 1×10^6 PBMCs/well in a 96-well round bottom tissue culture plate with TLR ligands (see Table 3 for working concentrations). Include wells for unstimulated and isotype controls.
- Incubate plates in a 37°C CO_2 tissue culture incubator for 12 hrs.
- Start processing for surface staining using procedures in 2.1.1., and either the DC (Table 1, Panel B) or monocyte (Table 2, Panel B) co-stimulatory molecule antibody panels. Fix samples using 3% paraformaldehyde and store in freezing medium at –80°C until analysis as discussed above in 2.1.1.

3.2. *TLR ligand induced cytokine expression in monocytes and dendritic cells*

3.2.1 *In vitro stimulation of monocytes for cytokine production*

- Incubate 1×10^6 freshly isolated PBMCs/well in a round bottom 96-well plate with TLR ligands (see Table 3 for working concentrations), including unstimulated and isotype controls.
- Dilute Brefeldin A stock from the manufacturer (Golgi Plug, BD Biosciences) 1:100 in warm complete RPMI medium just before use. BD GolgiPlug is stored at 4°C and must be warmed to room temperature prior to use.
- Incubate plates at 37°C in a CO_2 incubator for 12 hrs, then add 20 μL diluted Brefeldin A to each well to trap cytokines within cells. Incubate for an additional 6 hrs at 37°C.

Table 2.   Flow Cytometry Antibody Panel for Human Monocytes.

Fluorochrome	FITC	PE	PE-Texas Red	PerCP	Alexafluor700	PE-Cy7	APC	APC-Cy7	Pacific Blue
A. Monocyte Intracellular Cytokine Panel	Anti-IL-6 (MQ2-13A5)	Anti-IL-12p70 (20C2)	Anti-CD14 (TuK4)		Anti-TNFα (MAB11)	Anti-CD16 (3G8)	Anti-CD11c (B-Ly6)	Anti-CD11b (ICRF44)	Anti-IL-10 (JES3-9D7)
B. Monocyte Co-Stimulatory Panel	Anti-CD86 2331 (FUN – 1)	Anti-SLAM CD150 A12(7D4)	Anti-CD14 (TuK4)		Anti-CD80 (L307.4)	Anti-CD16 (3G8)	Anti-CD11c (B-Ly6)	Anti-CD11b (ICRF44)	Anti-CD62L (DREG-56)
C. Monocyte TLR Panel-I	Anti-TLR5 (85B152.5)	Anti-TLR1 (3.9)	Anti-CD14 (TuK4)	Anti-TLR2 (383936)	Anti-TLR6 (TLR6.127)	Anti-CD16 (3G8)	Anti-CD11c (B-Ly6)	Anti-CD11b (ICRF44)	
D. Monocyte TLR Panel-II	Anti-TLR8 (T2.5)	Anti-TLR3 (TLR3.7)	Anti-CD14 (TuK4)		Anti-TLR4 (HTA125)	Anti-CD16 (3G8)	Anti-CD11c (B-Ly6)	Anti-CD11b (ICRF44)	

- Following incubation, use the surface staining protocol from 2.1.1 and the monocyte intracellular staining panel (Table 2, panel A), excluding anti-cytokine antibodies.
- Fix samples using 3% Paraformaldehyde and store in freezing medium until acquisition as discussed in 2.1.1.
- On the day of analysis, perform intracellular staining as in 2.1.2, using antibodies recognizing IL-6, IL-12p70, TNFα and IL-10 (Table 2, panel A), or other cytokines as desired.

3.2.2 *In-vitro stimulation of dendritic cells for cytokine production*

- Incubate 1×10^6 freshly isolated PBMCs/well in a round-bottom 96- well plate with TLR ligands (see Table 3), including wells for unstimulated and isotype controls.
- Incubate plates at 37°C in a CO_2 tissue culture incubator for 2 hrs.
- After 2 hrs of incubation add of 20 μL Brefeldin A/well, diluted as in 3.2.1.
- Incubate the plates after addition of Brefeldin A for an additional 4 hrs.
- After 6 hrs of total incubation, prepare for flow cytometry staining using the protocol in 2.1.1. and the DC intracellular staining panel (Table 1, panel A), excluding anti-cytokine antibodies.

Table 3. Synthetic Toll-like Receptor Ligands for Human Peripheral Blood Stimulation.

TLR	Ligand	Working Concentrations μg/mL	Ligand Type
TLR1/2	Pam3CSK4	10	Tri-acylated lipoprotein
TLR2/6	LTA	2	Lipoteichoic Acid
TLR3	Poly (I:C)	10	HMW ds RNA
TLR4	LPS	1	Lipopolysaccharide
TLR5	Flagellin	5	Flagellin
TLR7/8	R848	0.5	SS RNA
TLR9	CpG	3	CPG ODN-B (ODN2006)

- Fix samples using 3% Paraformaldehyde and store in freezing medium until acquisition as discussed in 2.1.1.
- On the day of acquisition, perform intracellular staining following permeabilization method as discussed in 2.1.2 for IFNα, IL-6, IL-12p70 and TNFα (Table 1, panel A), or others, as desired.

4. Comments on Analysis

As described, we carry out all stimulations and surface staining on freshly isolated PBMCs, followed by fixing and freezing at −80°C. This allows us to collect and process fresh samples on different days, but complete intracellular staining and carry out data acquisition of all samples together. We have used Becton Dickinson Fortessa instruments using the Cytometer Setup and Tracking (CS&T) system for instrument setup. The report generated after successful bead tracking provides values for detector efficiency, optical background and electronic noise. By using the CS&T report, we take a rational approach (discussed below) to set up the instrument for our cell types of interest and available fluorochromes used in a specific experiment.

4.1 *Rational Approach*

- Prepare a tube of unstained cells as used in the experiment. For instance, if the experiment requires stimulation to determine an output such as cytokine production or co-stimulatory protein expression, use only activated unstained cells.
- Fix cells appropriately, if the experimental samples involve fixation.
- Prepare aliquots of single-color stained cells for each fluorochrome in the panel designed for a given experiment.
- Use the exact fluorochrome as indicated. For example, FITC cannot be replaced by Alexafluor 488 and APC Cy7 cannot be replaced by APC eF780.
- Run unstained cells to determine auto fluorescence.
- Run single-color stained cells to determine the positions of positive and negative peaks.

- Adjust the photomultiplier tube (PMT) voltage so that the positive peak is at half of the linear maximum range listed in the CS&T report.
- Similarly, adjust the PMT voltage so that the negative peak is above two and half times the electronic noise SD_{en} (rSD > 2.5 × SD_{en}); the SD_{en} value is available in the CS&T report.
- Repeat the process to assign proper peak positions for each fluorochrome used in the panel.
- While setting up all the parameters, if the positive peak for any parameter goes beyond the maximum limit, then reduce the PMT voltage for that particular parameter to keep the peak within range (i.e. half of the linear maximum range).
- Once the optimum values are determined for a given experiment, they can be saved under "APPLICATION SETTINGS" in the BD FACSDiva software and can be invoked multiple times for that particular experiment.
- Application settings are necessary for every experiment if the cell types are different or a new fluorochrome is used — such as if fresh cells are used instead of fixed cells, or FITC-conjugated antibody is replaced with Alexafluor 488-conjugated antibody.
- To obtain reproducible results, these steps need to be repeated whenever the instrument undergoes laser alignments or filter changes.

References

1. van Duin D, Medzhitov R, Shaw AC. (2006) Triggering TLR signaling in vaccination. *Trends Immunol* **27**: 49–55.
2. Shaw AC, Panda A, Joshi SR *et al.* (2011) Dysregulation of human Toll-like receptor function in aging. *Ageing Res Rev* **10**: 346–353.
3. Takeda K, Akira S. (2015) Toll-like receptors. *Curr Protoc Immunol* **109**: 14.12.1–14.12.10.
4. Shaw AC, Goldstein DR, Montgomery RR. (2013) Age-dependent dysregulation of innate immunity. *Nat Rev Immunol* **13**: 875–887.
5. Janeway CA, Jr., Medzhitov R. (1999) Lipoproteins take their toll on the host. *Curr Biol* **9**: R879–R882.

6. van Duin D, Mohanty S, Thomas V *et al.* (2007) Age-associated defect in human TLR-1/2 function. *J Immunol* **178**: 970–975.
7. Panda A, Qian F, Mohanty S *et al.* (2010) Age-associated decrease in TLR function in primary human dendritic cells predicts influenza vaccine response. *J Immunol* **184**: 2518–2527.
8. Rasmussen SM, Bilgrau AE, Schmitz A *et al.* (2015) Stable phenotype of B-cell subsets following cryopreservation and thawing of normal human lymphocytes stored in a tissue biobank. *Cytometry B Clin Cytom* **88**: 40–49.
9. Weinberg A, Song LY, Wilkening C *et al.* (2009) Optimization and limitations of use of cryopreserved peripheral blood mononuclear cells for functional and phenotypic T-cell characterization. *Clin Vaccine Immunol* **16**: 1176–1186.
10. Blimkie D, Fortuno ES, 3rd, Yan H *et al.* (2011) Variables to be controlled in the assessment of blood innate immune responses to Toll-like receptor stimulation. *J Immunol Methods* **366**: 89–99.
11. Vollmer J, Weeratna R, Payette P *et al.* (2004) Characterization of three CpG oligodeoxynucleotide classes with distinct immunostimulatory activities. *Eur J Immunol* **34**: 251–262.

Chapter 2

Dissecting Complex Cellular Systems with High Dimensional Single Cell Mass Cytometry

Mikael Roussel[*,†,‡], *Allison R. Greenplate*[†]
and Jonathan M. Irish[†,‡,§,¶]

[*]*Hematology Laboratory and INSERM U917*
University Hospital, Rennes, France
[†]*Department of Pathology, Microbiology and Immunology*
Vanderbilt University School of Medicine, Nashville, TN, USA
[‡]*Department of Cancer Biology and Vanderbilt-Ingram*
Cancer Center, Vanderbilt University School of Medicine
Nashville, TN, USA
[§]*Vanderbilt University School of Medicine, 740B Preston*
Building, 2220 Pierce Avenue, Nashville, TN, USA
[¶]*jonathan.irish@vanderbilt.edu*

1. Introduction

Analysis at the single cell level is crucial for characterizing cells within complex, heterogeneous populations.[1,2] This chapter explores the history of single cell biology in the mononuclear phagocyte system and the contributions that new measurement and analysis tools

---

[¶]Corresponding author.

have made to describe these cells. Mononuclear phagocytes represent a particular challenge due to the large number of phenotypes that these cells adopt after maturation and infiltration into tissues. The ability of fluorescent flow cytometry to interrogate individual cells has driven the modern era of immunology and revealed the details of the innate immune system. However, spectral overlap constrains the number of parameters that are routinely measured. Mass cytometry is a high dimensional, quantitative, single cell flow cytometry approach that uses time of flight mass spectrometry as a detection tool.[3] By coupling antibodies to metal isotopes instead of fluorophores, mass cytometry using a Cytometry by Time of Flight (CyTOF) instrument circumvents limitations of fluorescent spectral overlap and endogenous cellular auto-fluorescence and enables simultaneous measurement of more than 35 cellular features. The quantitative, high throughput nature of mass cytometry has sparked a new era of comprehensive single cell biology studies of complex cellular systems.[4,5]

2. The Mononuclear Phagocyte System

2.1. *Early observations*

From the start, phagocytes were described using light microscopy as cells that could engulf other particles, such as pathogens (Fig.1). With the development of the cluster of differentiation system,[6] great strides were made in tracking and characterizing heterogeneous populations of mononuclear phagocytes.[7–9] A consensus emerged that monocytes could be divided into key subsets using surface markers, and three functional subsets were termed "classical", "intermediate", and "non-classical" based on contrasting expression of CD14 and CD16 (Fig.1).[10,11] However, these definitions were recognized as incomplete from the start and much effort has been put into identifying additional markers of cell subsets. For example, CCR2 and Tie-2 have been associated with contrasting polarization roles in down- and up-regulating the inflammatory phenotype, respectively.[12,13]

2.2. *Phenotype and function versus lineage identity*

In contrast with adaptive immune cells, mononuclear phagocytes are not defined by lineage-restricted cell surface signaling complexes. Intracellular signaling and signals received do play crucial roles in polarization, but signaling receptors are not thought to specifically define monocyte or macrophage subsets. Many markers that are closely associated with myeloid cell function and identity are expressed on other cells, including CD14 on endothelial and epithelial cells[14] and HLA-DR on B cells, activated T cells, and cancer cells of diverse origins, including neural origin melanoma cells.[15] The cells known as "mononuclear phagocytes" include monocytes, dendritic cells, and tissue resident macrophages.[16] The main difference between this term and "myeloid cells" is the implication of lineage origin; some mononuclear phagocytes originate from outside the myeloid lineage.[8] The apparent convergent development of mononuclear phagocytes from different lineages raises the fundamental question: are cell populations defined by lineage ancestry or by phenotype and function?[16,17]

2.3. *Complexity in biology and nomenclature: polarization subsets, M1 versus M2, TAMs, and MDSCs*

Reference panels for blood monocyte evaluation include three core markers, HLA-DR, CD14, and CD16.[10,18–20] Other commonly used markers include CD13, CD33, and CD11b, which have been proposed to decipher monocyte maturation and differentiation in bone marrow and blood.[21] During infection, monocytes migrate through tissues where they can differentiate in dendritic cells or macrophages.

Within tissues, CD68 and CD163 are frequently proposed to characterize macrophage types.[22] These cell populations, which are involved in tissue homeostasis and host defense, were historically split into classically-activated or "M1" and alternatively-activated or "M2". M1 polarization occurs in response to IFN-γ or LPS stimulation and is associated with an increase of inflammatory cytokines and tumoricidal capabilities. In contrast, IL-4, IL-10, or IL-13

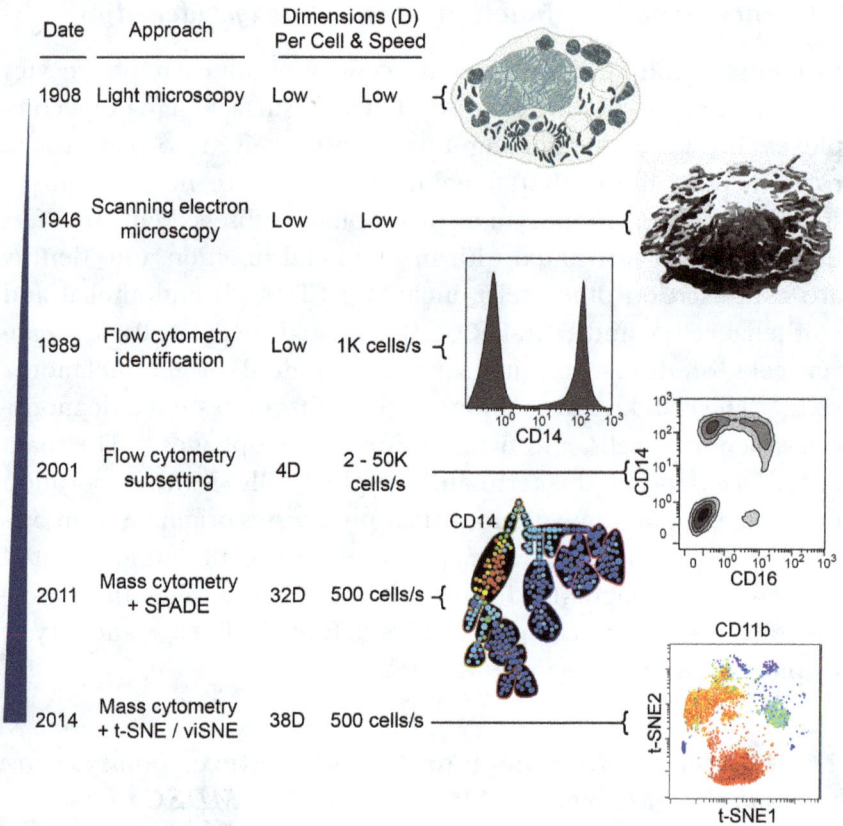

Date	Approach	Dimensions (D) Per Cell & Speed	
1908	Light microscopy	Low	Low
1946	Scanning electron microscopy	Low	Low
1989	Flow cytometry identification	Low	1K cells/s
2001	Flow cytometry subsetting	4D	2 - 50K cells/s
2011	Mass cytometry + SPADE	32D	500 cells/s
2014	Mass cytometry + t-SNE / viSNE	38D	500 cells/s

Fig. 1.   Macrophages through the Ages: From Microscopy to Mass Cytometry. The first observations of mononuclear phagocytes were made by microscopy in 1908 and noted the presence of lysosomes and the engulfment and destruction of bacteria.[23,24] In the 1940s, electron microscopy provided a clear view of macrophage shape and the pseudopods that seek extracellular particles and help direct macrophage movement.[25] In 1983, the 3C10 antibody for CD14 and other myeloid lineage antibodies were developed.[26] By 1989, CD14 had become standard in flow cytometry studies of mononuclear phagocytes, but multidimensional analysis was not yet widespread. At the start of the new millennium, multidimensional analysis was becoming mainstream and 2-laser cytometers capable of routine 4D analysis were widespread. Multidimensional dimensional analysis revealed additional complexity within cells known to be mononuclear phagocytes, and terms like 'classical' $CD14^{pos}$ $CD16^{neg}$ cells, 'intermediate' $CD14^{pos}$ $CD16^{pos}$ cells, and 'non-classical' $CD14^{neg}$ $CD16^{pos}$ were developed based on apparent clusters in 2D flow cytometry.[10] In 2011, mass cytometry characterized all human bone marrow cells with a single 32-antibody panel.[3] Key to this work was the use of the unsupervised clustering tool

◄

Fig. 1. (*Figure on facing page*) SPADE, which infers a phenotypic tree of cell population clusters identified in high-dimensional data.[27] In 2014, mass cytometry measurements of eight mouse tissues using a 38-antibody panel developed for the myeloid system created a comprehensive reference map of the myeloid system.[28] This study was also powered by unsupervised dimensionality reduction tools developed for machine learning, including t-SNE and ISOMAP. The 2011 and 2014 studies both relied on unsupervised tools that revealed *cyto incognito* — hidden cells with unexpected phenotypes that would have been overlooked in traditional analysis.[1]

stimulation polarizes macrophages to an M2 associated with tissue repair, pro-angiogenesis, and a lack of effective tumor immunity. In fact, these two types capture functions that are part of a wide spectrum of overlapping polarization states that depend largely on programming from external stimuli.[22,29]

Tumor-associated myeloid cells can exhibit immunosuppressive properties mediated by soluble or membrane-bound factors such as TNFα, IL-10 or CD163. The identification of the mechanisms responsible for the selective recruitment and acquisition of an immunosuppressive phenotype is the subject of intense research.[30] Indeed, manipulating locally the immune microenvironment by blocking recruitment of precursors or altering the suppressor cells *in situ* may improve antitumor immune responses. One approach to this is to selectively reprogram macrophages to promote anti-tumor immunity.[31] In most cancers, macrophages are the most abundant tumor-infiltrating immune cells. Tumor-associated macrophages (TAM) are immunosuppressive and often exhibit M2 characteristics that include expression of immunoregulatory molecules (e.g. B7-H4), an IL-12^{low} IL-10^{high} secretion profile, and the capacity to inhibit effector T cell functions. TAMs thus represent an attractive target for immunotherapies directed at the tumor microenvironment. An improved understanding of macrophage phenotype would greatly aid in selectively targeting functionally distinct macrophage subsets.

Human myeloid derived suppressor cells (MDSCs) also exhibit anti-tumor activity, but are distinct from TAMs in that they include myeloid cells other than macrophages and function in healthy regulatory contexts and diseases other than cancer. MDSCs are generally divided into two groups: monocytic and granulocytic MDSCs.[32,33] MDSCs suppress both innate and adaptive anti-tumor immunity through mechanisms including: T_{reg} development, deprivation of essential amino acids, and release of oxidizing molecules.[34] In cancer, MDSCs are recruited and activated by soluble factors secreted both by cancerous and host stromal cells within the tumor microenvironment.[32,34] MDSCs are a major contributor to immune dysfunction of patients with significant solid tumor burdens.[32] The relationships between MDSCs observed in cancer and TAMs are not well understood and it is not known to what extent cancers of the myeloid lineage depend on distinct properties of TAMs and MDSCs.

While phenotypically distinct subsets of monocytes, macrophages, TAMs, and MDSCs have focused functional roles,[35–38] it remains clear that the mononuclear phagocyte system is a broad *continuum* of phenotypes and that the classification systems are partially overlapping. Recently, mouse innate immune cells were comprehensively characterized using a 38-antibody mass cytometry panel.[28] A comparable high dimensional mass cytometry and machine learning study of human mononuclear phagocytes could help to bring clarity to this system.

3. Revisiting the Mononuclear Phagocyte System with High Dimensional Single Cell Analysis

3.1. *Mass cytometry and machine learning*

The ~5 to 10- fold increase in the number of per-cell features routinely measured in mass cytometry experiments has resulted in a massive increase in data complexity and revealed unexpected phenotypic patterns on well-studied cell populations. Traditionally, flow cytometry data are analyzed manually using biaxial gates. This type of analysis is highly susceptible to bias and requires prior knowledge of phenotype — a major limitation in the setting of disease. Additionally, traditional biaxial gating is impractical, since a routine CyTOF experiment measuring 32 features would produce 496 possible biaxial combinations that might still incompletely represent multidimensional phenotypic continuums. This problem has been extensively researched in the field of machine learning, where multidimensional analysis, clustering, and trajectory analysis are common themes.[11,27,28,39–42] Advances in computational biology have helped power high dimensional single cell biology and have provided researchers with powerful dimensionality reduction and cell classification tools.

SPADE and viSNE are unsupervised tools that can reduce high-dimensional data to a 2D map. Spanning-tree progression analysis of density-normalized events (SPADE) and is an unsupervised machine learning tool that clusters cells into nodes based on selected features.[7–9,27] SPADE clusters cells into groups (represented by a circle) and organizes those groups into a hierarchy of related phenotypes.[5,10,11,39] Statistics are displayed for each group.[12,13,27] More recently, Amir and colleagues adapted the t-stochastic neighbor embedding (t-SNE) algorithm to create a tool called visualization of t-SNE (viSNE).[16,40] viSNE software arranges single cells on a 2D 'map' based on their phenotypic similarity to each other in high-dimensional space, where 'islands' on the map are comprised of phenotypically similar cells.[10,18–20,40] A heatmap where color represents protein expression on cells can be used to characterizes the results of both viSNE and SPADE. Cellular abundance is represented by the size of the population in SPADE, whereas viSNE represents cellular abundance with a separate density map, comparable to that of a contour plot. Both of these tools analyze data in an unsupervised fashion, reducing individual bias and allowing for identification of cell populations with unusual or novel phenotypes.[1,21] The use of tools like SPADE and viSNE in combination with mass cytometry has resulted in better characterization of many cell types, including the cells of the myeloid system.[22,28]

3.2. *Mass cytometry's contributions to myeloid biology*

Recent papers have shed light on the myeloid compartment by using mass cytometry to characterize human and mice cell populations.[28,35–38,41,42] Bodenmiller *et al.* profiled the dynamic response and crosstalk among immune cells from 14 blood populations.[28,41] Altogether, on 14 non-overlapping PBMC subsets, 12 phosphorylated proteins were analyzed upon 12 different stimuli at 8 time points. This resulted in more than 18,000 conditions tested per sample. After 15 m in of LPS stimulation, NFκB phosphorylation was activated on a subset of monocytes expressing CD14^{pos} HLA-DR^{int}. Of note, at 60 m in of stimulation, STAT3 became phosphorylated in CD4^{high} T-cells (which are not thought to be able to directly respond to LPS as they lack CD14/TLR4), suggesting intercellular crosstalk.[32,33,41] The same approach was applied to analyzed the clinical relevance of surgery-induced immune perturbations.[34,42] After surgery, a specific CD33^{pos} CD11b^{pos} CD14^{pos} HLA-DR^{low} subset of monocyte was found expanded with differential phosphorylation of STAT4, CREB, and NFκB.[32,42] Interestingly, single cell network profiles were correlated with the patient's clinical recovery.[1,2,42] Thus, these concordant results show the potential of CyTOF to analyze a large number of subsets arising from a heterogeneous population such as myeloid cells, as well as analyze a timecourse response with a large number of conditions.

Myeloid complexity also results from the tissue specialization of monocytes into various populations of macrophages.[3,43,44] To decipher the mouse myeloid system across different tissues, Becher *et al.* used a 38-antibody panel associated with computational tools to build a framework of reference (Fig.1).[4,5,28] By using an automated computational method for population identification, they observed known populations of tissue-resident macrophages. Strikingly, innate immune cell subsets that were not expected to be well defined by their panel turned out to be phenotypically distinct, even though the markers were chosen specifically for myeloid cells. These results point out both an advantage of a high dimensional single cell approach and the fact that much of our knowledge of protein expression is based on focused analysis panels and gating that restricts the view to known, well-described cell populations. Altogether, these results highlight the potential benefits by using CyTOF for in depth studies on the mononuclear phagocyte system heterogeneity.

4. Future Directions

Over the past 100 years our knowledge of the mononuclear phagocyte system has expanded in tandem with improvements in measurement tools (Fig.1). Flow cytometry has recently made a leap forward due to the combination of machine learning tools and high dimensional mass cytometry. Critically, this advance may help to resolve the differences in the field around population identity, especially within the area of suppressor cell phenotypes that may represent different descriptions of a largely overlapping population base. Alternatively, the increase in makers and sensitivity for rare cell populations defined by multiple markers may further fracture the identities of mononuclear phagocytes into more and more functionally distinct subsets. Either way, it seems we stand at the start of a new era where population identities within complex cellular systems can be automatically defined, quantified and compared.

References

1. Irish JM. (2014) Beyond the age of cellular discovery. *Nat Immunol* **15(12)**:1095–1097.
2. Irish JM, Doxie DB. (2014) High-dimensional single-cell cancer biology. *Curr Top Microbiol Immunol* **377**:1–21.
3. Bendall SC, Simonds EF, Qiu P *et al.* (2011) Single-Cell Mass Cytometry of Differential Immune and Drug Responses Across a Human Hematopoietic Continuum. *Science* **332(6030)**:687–696.
4. Di Palma S, Bodenmiller B. (2015) Unraveling cell populations in tumors by single-cell mass cytometry. *Curr Op Biotechnol.* **31**:122–129
5. Bendall SC, Davis KL, Amir E-AD *et al.* (2014) Single-Cell Trajectory DetectionUncovers Progression and Regulatory Coordination in Human B Cell Development. *Cell* **157(3)**:714–725.
6. Zola H, Swart B, Banham A *et al.* (2007) CD molecules 2006 — human cell differentiation molecules. *J. Immunol. Methods* **319(1–2)**:1–5.
7. Gordon S, Taylor PR. (2005) Monocyte and macrophage heterogeneity. *Nat Rev Immunol* **5(12)**:953–964.
8. Jenkins SJ, Hume DA. (2014) Homeostasis in the mononuclear phagocyte system. *Trends Immunol* **35(8)**:358–367.

9. Geissmann F, Gordon S, Hume DA. (2010) Unravelling mononuclear phagocyte heterogeneity. *Nat Rev Immunol* **10(6)**:453–460.

10. Ziegler-Heitbrock L, Ancuta P, Crowe S *et al.* (2010) Nomenclature of monocytes and dendritic cells in blood. *Blood.* **116(16)**:e74–e80.

11. Cros J, Cagnard N, Woollard K *et al.* (2010) Human CD14dim Monocytes Patrol and Sense Nucleic Acids and Viruses via TLR7 and TLR8 Receptors. *Immunity.* **33(3)**:375–386.

12. Murdoch C, Tazzyman S, Webster S *et al.*(2007) Expression of Tie-2 by human monocytes and their responses to angiopoietin-2. *J Immunol* **178(11)**:7405–7411.

13. Serbina NV, Pamer EG. (2006) Monocyte emigration from bone marrow during bacterial infection requires signals mediated by chemokine receptor CCR2. *Nat Immunol* **7(3)**:311–317.

14. Jersmann HPA. (2005) Time to abandon dogma: CD14 is expressed by non-myeloid lineage cells. *Immunol Cell Biol* **83(5)**:462–467.

15. D'Alessandro G, Zardawi I, Grace J. (1987) Immunohistological evaluation of MHC class I and II antigen expression on nevi and melanoma: Relation to biology of melanoma. *Pathol* **19(4)**:339–346.

16. Geissmann F, Manz MG, Jung S *et al.* (2010) Development of Monocytes, Macrophages, and Dendritic Cells. *Science* **327(5966)**:656–661.

17. Guilliams M, Ginhoux F, Jakubzick C *et al.* (2014) Dendritic cells, monocytes and macrophages: A unified nomenclature based on ontogeny. *Nat Rev Immunol* **14(8)**:571–578.

18. Abeles RD, McPhail MJ, Sowter D *et al.* (2012) CD14, CD16 and HLA-DR reliably identifies human monocytes and their subsets in the context of pathologically reduced HLA-DR expression by CD14(hi)/CD16(neg) monocytes: Expansion of CD14(hi)/CD16(pos) and contraction of CD14(lo)/CD16(pos) monocytes in acute liver failure. *Cytometry A* **81(10)**:823–834.

19. Autissier P, Soulas C, Burdo TH *et al.* (2010) Evaluation of a 12-color flow cytometry panel to study lymphocyte, monocyte, and dendritic cell subsets in humans. *Cytometry A* **77(5)**:410–419.

20. Bocsi J, Melzer S, Dähnert I *et al.* (2014) OMIP-023: 10-Color, 13 antibody panel for in-depth phenotyping of human peripheral blood leukocytes. *Cytometry A* **85(9)**:781–784.

21. van Lochem EG, van der Velden VHJ *et al.* (2004) Immunophenotypic differentiation patterns of normal hematopoiesis in human bone marrow: Reference patterns for age-related changes and disease-induced shifts. *Cytometry A* **60B(1)**:1–13.

22. Biswas SK, Allavena P, Mantovani A. (2013) Tumor-associated macrophages: Functional diversity, clinical significance, and open questions. *Semin Immunopathol* **35(5)**:585–600.

23. Gordon S. (2008) Elie Metchnikoff: Father of natural immunity. *Eur J Immunol* **38(12)**:3257–3264.

24. Metchnikoff E. (1968) Lectures on the Comparative Pathology of Inflammation Delivered at the Pasteur Institute in 1891. Starling, F. A. and Starling, E. H. (transl.), Dover Publications Inc., New York.

25. Bessis M. (1948) Cytologie sanguine normale et pathologique. Paris: Masson & Cie.

26. Van Voorhis WC, Steinman RM, Hair LS *et al.* (1983) Specific anti-mononuclear phagocyte monoclonal antibodies. Application to the purification of dendritic cells and the tissue localization of macrophages. *J Exp Med* **158(1)**:126–145.

27. Qiu P, Simonds EF, Bendall SC *et al.* (2011) Extracting a cellular hierarchy from high-dimensional cytometry data with SPADE. *Nat Biotechnol* **29(10)**:886–891.

28. Becher B, Schlitzer A, Chen J *et al.* (2014) High-dimensional analysis of the murine myeloid cell system. *Nat Immunol* **15(12)**:1181–1189.

29. Xue J, Schmidt SV, Sander J *et al.* (2014) Transcriptome-Based Network Analysis Revealsa Spectrum Model of Human Macrophage Activation. *Immunity* **40(2)**:274–288.

30. Mantovani A, Locati M. (2013) Tumor-associated macrophages as a paradigm of macrophage plasticity, diversity, and polarization: Lessons and open questions. *Arterioscler Thromb Vasc Biol* **33(7)**:1478–1483.

31. Yu SS, Lau CM, Barham WJ *et al.* (2013) Macrophage-specific RNA interference targeting via "click," mannosylated polymeric micelles. *Mol Pharm* **10(3)**:975–987.

32. Gabrilovich DI, Nagaraj S. (2009) Myeloid-derived suppressor cells as regulators of the immune system. *Nat Rev Immunol* **9(3)**:162–174.

33. Peranzoni E, Zilio S, Marigo I *et al.* (2010) Myeloid-derived suppressor cell heterogeneity and subset definition. *Curr Opin Immunol* **22(2)**:238–244.

34. Gabrilovich DI, Ostrand-Rosenberg S, Bronte V. (2012) Coordinated regulation of myeloid cells by tumours. *Nat Rev Immunol* **12(4)**:253–268.

35. Ingersoll MA, Spanbroek R, Lottaz C *et al.* (2010) Comparison of gene expression profiles between human and mouse monocyte subsets. *Blood* **115(3)**:e10–e19.

36. Wong KL, Tai JJ-Y, Wong W-C *et al.* (2011) Gene expression profiling reveals the defining features of the classical, intermediate, and non-classical human monocyte subsets. *Blood* **118(5)**:e16–e31.

37. Zawada AM, Rogacev KS, Rotter B *et al.* (2011) SuperSAGE evidence for CD14++CD16+ monocytes as a third monocyte subset. *Blood* **118(12)**:e50–e61.

38. Martinez FO, Helming L, Milde R *et al.* (2013) Genetic programs expressed in resting and IL-4 alternatively activated mouse and human macrophages: Similarities and differences. *Blood* **121(9)**: e57–e69.

39. Horowitz A, Strauss-Albee DM, Leipold M *et al.* (2013) Genetic and environmental determinants of human NK cell diversity revealed by mass cytometry. *Science Transl Med* **5(208)**:208ra145.

40. Amir E-AD, Davis KL, Tadmor MD *et al.* (2013) viSNE enables visualization of high dimensional single-cell data and reveals phenotypic heterogeneity of leukemia. *Nat Biotechnol* **31(6)**:545–552.

41. Bodenmiller B, Zunder ER, Finck R *et al.* (2012) Multiplexed mass cytometry profiling of cellular states perturbed by small-molecule regulators. *Nat Biotechnol.* **30(9)**:857–866.

42. Gaudilliere B, Fragiadakis GK, Bruggner RV *et al.* (2014) Clinical recovery from surgery correlates with single-cell immune signatures. *Science Transl Med* **6(255)**:255ra131.

43. Hashimoto D, Miller J, Merad M. (2011) Dendritic cell and macrophage heterogeneity in vivo. *Immunity.* **35(3)**:323–335.

44. Ginhoux F, Jung S. (2014) Monocytes and macrophages: Developmental pathways and tissue homeostasis. *Nat Rev Immunol* **14(6)**: 392–404.

Chapter 3

CyTOF: Single Cell Mass Cytometry for Evaluation of Complex Innate Cellular Phenotypes

Dara M. Strauss-Albee* and Catherine A. Blish*,†,‡

*Stanford Immunology
†Department of Medicine, Stanford University School
of Medicine, Stanford, CA 94305 USA
‡cblish@stanford.edu

Summary

Single cell mass cytometry is a next-generation flow cytometry platform that allows for simultaneous assessment of 34+ parameters in a single sample. This is achieved by analysis on a CyTOF (Cytometry by Time-Of-Flight) mass spectrometer. CyTOF detects signals from antibodies conjugated to metal isotopes, in contrast to the fluorophores used in traditional fluorescence cytometry. This discrete detection mechanism spans a 4-log range of sensitivity without significant overlap among channels. Here we describe an approach using mass cytometry to evaluate the complex phenotype of natural

‡Corresponding author.

killer (NK) cells. The techniques described here could be extended to address a multitude of phenotypic cellular states.

1. Introduction

Flow cytometry has been a mainstay technique of immunological analysis since its inception in the 1960s. Its ability to quantify the expression of both surface and intracellular markers has been an invaluable asset in advancing the field. In this technique, a cell suspension is hydrodynamically focused into a single-cell stream, which is interrogated with a series of lasers. The signal from fluorophore-conjugated antibodies on each cell (surface or intracellular) is then quantified based on emission spectra.

However, the use of fluorescence to measure cellular phenotype and function also presents limitations. Overlapping emission spectra create a need for compensation across channels. Additionally, intrinsic cellular auto fluorescence can cause high background staining. Mass cytometry overcomes both of these limitations. By using antibodies coupled to rare metal isotopes, it detects discrete ion peaks which do not require compensation.[1-3] In addition, more than 40 isotopically pure metals are available, dramatically increasing the dimensionality from a typical 8–12 parameters to more than 34.[4,5]

While this increase in total number of parameters may seem modest, it corresponds to a much more impressive increase in the number of detectable cellular phenotypes. For instance, by Boolean analysis, a 12-color fluorescence cytometry panel can detect 2^{12} or 4096, distinct cellular populations. A similar Boolean analysis of a 34-parameter mass cytometry panel can detect 2^{34} (more than 17 billion) subpopulations. Thus, even before adding the complexity of functional analysis, the mass cytometry platform allows for a far more comprehensive evaluation of immunity than has been previously possible.

Pre-conjugated antibodies for mass cytometry have recently become available; however, reagents may still be limiting for deep phenotypic exploration of innate cell populations. Custom conjugation of antibodies is a straightforward process that allows a wider range of markers to be interrogated, while also granting the additional flexibility of custom panel design. Here, we describe a method to profile NK cell phenotype by custom conjugating antibodies, labeling cells, and analyzing on a CyTOF mass cytometer. This approach could be applied to many cell types by procuring and conjugating the appropriate antibodies.

2. Materials

All reagents must be free from heavy metal contaminants. Therefore, for mass cytometry experiments, no materials washed with soap should be used, as it can contain trace contaminants. Follow proper disposal techniques at all times. Store all materials according to manufacturer's recommendations.

2.1 *Antibody conjugation supplies*

(1) Maxpar® Metal labeling kits (DVS Sciences/Fluidigm Corporation).
(2) Purified monoclonal antibodies (see Table 1 for examples), purified IgG or polyclonal, must not have any carrier protein, otherwise special order is required. Sodium azide is acceptable.
(3) Centrifugal Filter Unit: 50 kDa Amicon Ultra — 500 μL V bottom (Millipore UFC505396) OR 30 kDa Amicon Ultra 500 μL V bottom (Millipore UFC503096).
(4) Centrifugal Filter Unit: 3 kDa Amicon Ultra — 500 μL V bottom (Millipore UFC503096).
(5) Aerosol Barrier (Filter) Pipette Tips.
(6) 0.5 M TCEP (Tris(2-carboxyethyl)phosphine): Bond-breaker TCEP Solution (Pierce #77720).
(7) Microcentrifuge, ideally 2 units.
(8) Heat block incubator or water bath at 37 ̊C.

Table 1.   Example Mass Cytometry Antibody Panel Information for Characterization of NK cells.

Isotope	Isotope Source	Antibody	Antibody Clone	Antibody Source
112Cd		CD3 Qdot	S4.1	Invitrogen
141Pr	DVS Sciences	CD27	O323	BioLegend
142Nd	DVS Sciences	CD19	SJ25-C1	Southern Biotech
143Nd	DVS Sciences	CD4	SK3	BioLegend
144Nd	DVS Sciences	CD8	SK1	BioLegend
145Nd	DVS Sciences	CD57	HCD57	BioLegend
146Nd	DVS Sciences	KIR2DL1/S1	EB6.B	Beckman Coulter
147Sm	DVS Sciences	TRAIL	RIK-2	BioLegend
148Nd	DVS Sciences	KIR2DL2/L3/S2	GL183	Beckman Coulter
149Sm	DVS Sciences	CD16	3G8	BioLegend
151Eu	DVS Sciences	KIR3DL1/S1	Z27	Beckman Coulter
153Eu	DVS Sciences	KIR2DS4	FES172	Beckman Coulter
154Gd	DVS Sciences	LILRB1	HPF1.4	Beckman Coulter
155Gd	Trace Sciences	NKp46	195314	R&D Systems
156Gd	DVS Sciences	NKG2D	1D11	BioLegend
157Gd	Trace Sciences	NKG2C	134591	R&D Systems
158Gd	DVS Sciences	2B4	2-69	BD Pharmingen
159Tb	DVS Sciences	CD33	WM53	BioLegend
160Gd	DVS Sciences	CD11b	ICRF44	BioLegend
161Dy	Trace Sciences	NKp30	P30-15	BioLegend
163Dy	Trace Sciences	KIR3DL1	DX9	BD Pharmingen
164Dy	DVS Sciences	NKp44	P44-8	BioLegend
165Ho	DVS Sciences	CD127	A019D5	BioLegend
166Er	DVS Sciences	KIR2DL1	143211	R&D Systems
167Er	DVS Sciences	CD94	DX22	BioLegend
169Tm	DVS Sciences	CCR7	2H4	BD Pharmingen
170Yb	DVS Sciences	KIR2DL3	180701	R&D Systems
171Yb	DVS Sciences	NKG2A	Z199.1.10	Beckman Coulter
172Yb	DVS Sciences	HLA-DR	L243	BioLegend
174Yb	DVS Sciences	CD56	NCAM16.2	BD Pharmingen
175Lu	DVS Sciences	KIR2DL5	UP-R1	BioLegend
195Pt	Enzo Life Sciences	Cisplatin live/ dead		

(9) PBS-based antibody stabilization solution (Candor Biosciences #131050).

(10) Nanodrop for protein quantification.

(11) Gd155 and Gd157 are available from Trace Sciences as a special order.

2.2 Mass cytometry labeling supplies

(1) Custom conjugated antibodies or conjugated antibodies purchased from DVS Sciences/Fluidigm Corporation.

(2) Thermo Scientific™ Nunc™ 96 Deep Well Plates, Polystyrene (Fisher Scientific 12-565-552). Conventional 96-well round bottom plates can be used, but require a larger number of lower volume washes, leading to reduced cell recovery.

(3) MilliQ dH2O. No water should have contact with beakers or bottles washed with soap.

(4) CyPBS: PBS without heavy metal contaminants, made from 10× PBS (Rockland Immunochemicals #MB-011) using MilliQ purified water, with no contact with glassware washed with soap.

(5) CyFACS buffer: 0.1% bovine serum albumin + 2 mM EDTA + 0.1% sodium azide in CyPBS. Filter solution with a 0.2 μM filter.

(6) Cisplatin Solution (Enzo Life Sciences #ALX-400-040-M250), diluted to 100 mM in DMSO.

(7) 0.1 μM spin filters (Millipore #UFC30VVOO).

(8) 16% Paraformaldehyde (Electron Microscopy Sciences #15710).

(9) Iridium interchelator solution (DVS Sciences #201192).

(10) RPMI-1640 Media (Hyclone/Thermo Scientific SH30096.01).

(11) Refrigerated centrifuge equipped with rotor for spinning 96 well plates.

(12) Aspirator with vacuum trap set-up.

2.3 CyTOF mass cytometry running supplies

(1) Ice bucket.

(2) Micropipettes.

(3) Normalization beads (DVS Sciences #201078).

3. Methods

Carry out all procedures at room temperature (RT) unless otherwise specified. Antibody conjugation protocols are well-described in the directions accompanying the kit. Labeling procedures are very similar to standard flow cytometry.

3.1 *Antibody conjugation using maxpar metal labeling kits*

Follow manufacturer's instructions for conjugation of antibodies. Custom-conjugating antibodies gives much more flexibility in panel design than would be available by purchasing pre-conjugated antibodies. An example of a panel to profile NK cell phenotype is shown in Table 1. This panel contains receptors to identify major cell lineages (such as CD3, CD4, CD8, CD19, CD56, C11b, and CD33) as well as markers for the major NK cell receptor families including the killer immunoglobulin-like receptors (e.g. KIR2DL1/S1, KIR2DL2/L3/S2), Fc receptors (CD16), natural cytotoxicity receptors (NKp30, NKp44, NKp46), C-type lectin-like receptors (NKG2A, NKG2C, NKG2D), and markers of maturity and differentiation (e.g. CD57, CD27, CCR7). To create a panel for other cells of interest, assign metals from available reagents and custom-conjugate additional markers as needed to remaining metal channels. Greater signal-to-noise ratio will be achieved on isotopes in the middle of the mass range. As a result, markers that readily distinguish expressing versus non-expressing cells should be reserved for channels at the extremes of the spectrum.

Special notes to supplement the kit instructions: When mixing the polymer with the L-buffer at the beginning of the protocol, mix very thoroughly. Be cautious to incorporate all the polymer as the polymer does not always reach the bottom of the tube during the recommended spin. It is also important not to pipette forcefully onto the filter when re-suspending the lanthanide-loaded polymer before conjugation with the partially reduced antibody. Instead, lift the pipette tip and make circular motions to gently re-suspend the polymer on the filter. As purified antibodies can have lot-to-lot variation, keep track of lot numbers for each conjugation. Further, use a single lot and conjugation for a set of linked experiments, whenever possible.

3.2. *Surface labeling of cells for mass cytometry*

(1) Prepare antibody staining cocktail. This is best performed using a spreadsheet to calculate the quantities. An example is shown in Table 2.

(2) Labeling is most conveniently performed in a deep well 96 well plate. As yields are much lower for mass cytometry than for flow cytometry (capture of ~30% compared to ~95% of events), 2 million cells per well is a good starting point, though the final number to use ultimately depends on the frequency of the cell population of interest. Incubate cells according to experimental design under investigation and then aliquot cells into wells and centrifuge plate at $750 \times g$ (typically ~1500 RPM on most tabletop units) for 10 m.

(3) While cells are pelleting, add 100 μL of the 100 mM cisplatin stock solution to 900 μL PBS to make a 10 mM cisplatin working solution.

(4) Working quickly, dilute the cisplatin working solution 1:200 in serum and antibiotic-free RPMI, for a final concentration of 50 μM. Re-suspend cells in 400 μL of this final stock.

(5) Incubate cells 1 m at RT.

(6) Quench with 400 μL of serum. Pipette up and down to mix thoroughly.

(7) Centrifuge plate at $750 \times g$ for 10 m at RT.

(8) After removing plate, set centrifuge to 4 °C in preparation for Step 13. Flick plate to remove supernatant. This is best accomplished by vigorously inverting the plate over a waste receptable, and blotting while still inverted on clean paper towels or other absorbent material.

(9) Spin antibody cocktail through a 0.1 μM Millipore spin filter for 3 m at 10,000 $\times$ g (~10,000 RPM on most microfuges) immediately before staining.

(10) Resupend cells (from step 7) in 50 μL of the antibody staining cocktail.

(11) Incubate 30 m on ice.

Table 2.   Example of Labeling Cocktail Antibody Solution.

Isotope	Antibody	Antibody concentration[1]	Date conj.[2]	Working concentration[3]	N+1[4]	Volume to add
Qdot - Cd	CD3	0.25 ul/test	—		11	2.75
141Pr	CD27	282	1/1/14	2.5	11	4.88
142Nd	CD19	347	1/1/14	5.00	11	7.93
143Nd	CD4	475	1/1/14	2.50	11	2.89
144Nd	CD8	433	1/1/14	0.63	11	0.79
145Nd	CD57	204	1/1/14	0.10	11	0.27
146Nd	KIR2DL1/S1	309	1/1/14	5.00	11	8.90
147Sm	TRAIL	310	1/1/14	0.20	11	0.35
148Nd	KIR2DL2/L3/S2	266	1/1/14	1.25	11	2.58
149Sm	CD16	445	1/1/14	0.20	11	0.25
151Eu	KIR3DL1/S1	327	1/1/14	0.63	11	1.05
153Eu	KIR2DS4	292	1/1/14	1.25	11	2.35
154Sm	LILRB1	393	1/1/14	2.50	11	3.50
155Gd	NKp46	474	1/1/14	2.50	11	2.90
156Gd	NKG2D	269	1/1/14	0.63	11	1.29
157Gd	NKG2C	249	1/1/14	1.25	11	2.76
158Gd	2B4	276	1/1/14	2.50	11	4.98

(Continued)

Table 2. (*Continued*)

Isotope	Antibody	Antibody concentration[1]	Date conj.[2]	Working concentration[3]	N+1[4]	Volume to add
159Tb	CD33	417	1/1/14	0.10	11	0.13
160Gd	CD11b	387	1/1/14	2.50	11	3.55
161Dy	NKp30	292	1/1/14	2.50	11	4.71
163Dy	KIR3DL1	493	1/1/14	1.25	11	1.39
164Dy	NKp44	289	1/1/14	0.30	11	0.57
165Ho	CD127	224	1/1/14	1.25	11	3.07
166Er	KIR2DL1	313	1/1/14	1.25	11	2.20
167Er	CD94	405	1/1/14	1.25	11	1.70
169Tm	CCR7	276	1/1/14	1.25	11	2.49
170Yb	KIR2DL3	255	1/1/14	2.50	11	5.39
171Yb	NKG2A	428	1/1/14	2.50	11	3.21
172Yb	HLA-DR	299	1/1/14	0.63	11	1.15
174Yb	CD56	294	1/1/14	0.63	11	1.17
175Lu	KIR2DL5	337	1/1/14	0.63	11	1.02
				Vol CyFACS[5]		**467.81**

[1]Antibody concentration determined by nanodrop, taking into account dilution in the antibody stabilization buffer

[2]We keep track of antibody lots, and do head-to-head testing of new conjugations

[3]Working concentration is determined by titration, typically starting at 10 mg/mL and testing serial 2-fold dilutions down to 0.16 mg/mL.

[4]N = number of tests. We make up enough for 1 additional sample

[5]Antibodies are diluted into CyFACS buffer solution for staining (see Section 2.2 of Materials)

(12) Add 700 μL CyFACS buffer to each well.

(13) Centrifuge plate at 750 × g for 10 m at 4°C.

(14) Flick plate to remove supernatant as described above.

(15) Add 800 μL CyFACS buffer to each well.

(16) Centrifuge at 750 × g (1500 RPM) for 10 m at 4°C.

(17) Prepare interchalator-PFA solution by diluting Ir-Interchelator 1:10,000 into paraformaldehyde (2% final concentration) in CyPBS.

(18) Re-suspend cells thoroughly in 100 μL of CyPBS containing Ir-Interchalator (1:10,000) + 2% PFA.

(19) Incubate overnight at 4°C.

(20) The next day (the same day as the cells will be run on the mass cytometer), add 600 μL of CyPBS buffer to each well.

(21) Centrifuge at 1000 × g (~2000 RPM) for 10 m at RT.

(22) Aspirate the supernatant, leaving 100 μL volume residual in the tube.

(23) Add 600 μL MilliQ water (metal-free) to each tube.

(24) Centrifuge at 1000 × g (~2000 RPM) for 10 m at RT.

(25) Aspirate the supernatant, leaving 100 μL residual in each tube.

(26) Repeat steps 23–25 twice for a total of three washes in MilliQ water.

(27) Re-suspend cells in the residual 100 μL at after the final wash.

(28) Run on mass cytometer.

3.3 *Running samples on a CyTOF mass cytometer*

The protocol for running samples has been covered elsewhere (see ref 6). It is important to note that when looking for rare cell populations, higher numbers of total cells will need to be run through the mass cytometer. Additionally, although the maximum collection rate is estimated at 1000 cells/s, to ensure the avoidance of doublets and nebulizer clots, it is prudent to use low run speeds (300 cells/s or lower). For assay normalization within and among runs, normalization beads (DVS Sciences #201078) should be used as described in ref 7.

4. Notes

(1) Follow notes/warnings on the antibody conjugation kit instructions, including the need to equilibrate polymer to RT to avoid moisture condensation.

(2) Conjugation with IgM is not supported by DVS/Fluidigm. However, we have found that IgM conjugations (e.g. CD57) have been successful, but with lower yields.

(3) As with fluorescence cytometry, protocols can be adapted for intracellular staining.

(4) Qdot antibodies have a cadmium core and can be detected by mass cytometry. However, this channel should be reserved for very bright antibodies (e.g. HLA-DR, CD3) or as a dump channel only. They can be used off-the-shelf following titration to determine optimal staining.

(5) Currently, DVS Sciences does not sell kits to conjugate to Gd155 and Gd157. These metals can be procured as 92%+ purity from Trace Sciences, and conjugated with polymer from Maxpar kits.

(6) Early protocols used DOTA-maleimide as a live-dead stain;[3,8–10] however, we find that cisplatin gives excellent resolution as a live dead marker[11] with the added benefit of preserving the In-115 channel to be used for other markers.

(7) All antibodies should be titrated to determine optimal concentration. A typical starting point is 10 mg/mL, with 6 serial 2-fold dilutions to test the range between 0.16 mg/mL and 10 mg/mL.

(8) It is straightforward and efficient to conjugate multiple antibodies at once. We have found that 8 conjugations provides an ideal balance of efficiency and precise incubation time.

(9) The MaxPar kits recommend using a 50 kDa filter, but we have also had success with 30 kDa units.

(10) Analysis of cytometry data is complex. A variety of tools are available for analysis, including SPADE,[12] Citrus,[13] viSNE,[14] Boolean analyses[9,10] and partially supervised gating and clustering algorithms[15] and new tools are released with some

regularity. Discussion of analytics is outside the scope of this chapter, but has been summarized elsewhere.[4,5]

Acknowledgements

DSA is supported by a Ruth L. Kirschstein National Research Service Award 1F31AI118469-01. Both authors are supported by a NIH Director's New Innovator Award DP2AI112193 (to CAB).

References

1. Ornatsky OI, Kinach R, Bandura DR *et al.* (2008) Development of analytical methods for multiplex bio-assay with inductively coupled plasma mass spectrometry. *J Anal At Spectrom* **23(4)**:463.
2. Ornatsky O, Baranov VI, Bandura DR *et al.* (2006) Multiple cellular antigen detection by ICP-MS. *J Immunol Methods* **308(1–2)**:68–76.
3. Bendall SC, Simonds EF, Qiu P *et al.* (2011) Single-cell mass cytometry of differential immune and drug responses across a human hematopoietic continuum. *Science* **332(6030)**:687–696.
4. Bendall SC, Nolan GP, Roederer M *et al.* (2012) A deep profiler's guide to cytometry. *Trends Immunol* **33(7)**:323–332.
5. Bjornson ZB, Nolan GP, Fantl WJ. (2013) Single-cell mass cytometry for analysis of immune system functional states. *Curr Opin Immunol* **25(4)**:484–494.
6. Leipold MD, Maecker HT. (2012) Mass cytometry: Protocol for daily tuning and running cell samples on a CyTOF mass cytometer. *JoVE* (69).
7. Finck R, Simonds EF, Jager A *et al.* (2013) Normalization of mass cytometry data with bead standards. *Cytometry* **83A(5)**:483–494.
8. Newell EW, Sigal N, Bendall SC *et al.* (2012) Cytometry by time-of-flight shows combinatorial cytokine expression and virus-specific cell niches within a continuum of CD8+ T cell phenotypes. *Immunity* **36(1)**: 142–152.
9. Horowitz A, Strauss-Albee DM, Leipold M *et al.* (2013) Genetic and environmental determinants of human NK cell diversity revealed by mass cytometry. *Sci Transl Med* **5(208)**:208ra145–5.
10. Strauss-Albee DM, Horowitz A, Parham P *et al.* (2014) Coordinated regulation of NK receptor expression in the maturing human immune system. *J Immunol.* **193(10)**:4871–4879.

11. Fienberg HG, Simonds EF, Fantl WJ *et al.* (2012) A platinum-based covalent viability reagent for single-cell mass cytometry. *Cytometry* **81A(6)**:467–475.

12. Qiu P, Simonds EF, Bendall SC *et al.* (2011) Extracting a cellular hierarchy from high-dimensional cytometry data with SPADE. *Nat Biotechnol* **29(10)**:886–891.

13. Bruggner RV, Bodenmiller B, Dill DL *et al.* (2014) Automated identification of stratifying signatures in cellular subpopulations. Proc Natl Acad Sci **111(26)**:E2770–E2777.

14. Amir E-AD, Davis KL, Tadmor MD *et al.* (2013) viSNE enables visualization of high dimensional single-cell data and reveals phenotypic heterogeneity of leukemia. *Nat Biotechnol* **31(6)**:545–552.

15. Finak G, Frelinger J, Jiang W *et al.* (2014) OpenCyto: An open source infrastructure for scalable, robust, reproducible, and automated, end-to-end flow cytometry data analysis. *PLoS Comput Biol* **10(8)**:e1003806.

Chapter 4

High-Throughput Secretomic Analysis of Single Cells to Assess Functional Cellular Heterogeneity

Kathryn Miller-Jensen[*,†] *and Rong Fan*[*,‡]

*Department of Biomedical Engineering, Yale University
New Haven, CT 06511 USA
†kathryn.miller-jensen@yale.edu
‡rong.fan@yale.edu

Summary

High throughput and highly multiplexed tools for secretion analysis are critical for interrogating innate immune system function. However, conventional assays that measure responses in cell populations obscure subsets of innate immune cells that may have distinct secretory functions, and do not provide information about natural variation in secretion responses that may be important for understanding mechanisms. Here we present a new microfluidic immunoassay for high throughput, multiplexed measurements of secretion in single living cells. This micro-well technique can be used to measure up to 16 cytokines simultaneously in up to 1000 cells. It does not require a sophisticated microfluidic control system to operate and

---

[†]Corresponding author.

thus can be implemented in most immunology labs with some basic microfluidic materials and equipment.

1. Introduction

Proteins secreted from innate immune cells play key roles in the regulation of the immune response to infection and injury.[1] Secretion of cytokines and other soluble factors can be initiated by immunoglobulin or complement receptor-mediated signaling or by pathogens via pattern recognition receptors (PRRs). The cascades of cytokines released by innate immune cells permit cell-to-cell communication of signals that positively and negatively regulate inflammatory responses.[2] Secreted cytokines also mediate recruitment and activation of T lymphocytes and other cells to mount adaptive immune responses. For these reasons, high throughput and highly multiplexed tools for secretion analysis are critical for interrogating immune system function.

The primary technique used to measure proteins secreted from cells is the enzyme-linked immunosorbent assay (ELISA). Traditional ELISAs measure secretion from populations of cells and are generally limited to detection of one protein target per assay. New technologies (such as Luminex bead-based assays) have increased multiplexing and throughput. However, because these measurements are population averages, it is impossible to identify sub-populations of innate immune cells that have distinct functionality, or to measure natural variation in secretion that may be functionally relevant.[3,4] Population averages also obscure co-regulation of genes or proteins, causal relationships between events and many other biological functions. For these reasons, there has been a significant effort to develop assays to interrogate signaling and phenotypic functions in single cells in the immune system.[5,6]

There are some established methods for measuring secreted proteins in single cells, including enzyme-linked immunoSpot (ELISpot) assay and intracellular cytokine staining (ICS)[7] by flow cytometry. However, each presents some disadvantages. ELISpot is a labor-intensive assay that generally requires long incubation times for secretion and has limited multiplexing capability. ICS has high throughput (typically thousands of cells are analyzed) and it can be multiplexed as with any flow cytometry assay, but the need to use a protein secretion inhibitor (typically brefeldin A or monensin) alters the functionality of the cell. A further disadvantage is after an ELISpot or ICS assay, the cells cannot be recovered for further analysis.

Here we present a new microfluidic assay for high throughput, multiplexed measurements in single living cells.[8] Microfluidic systems are designed to handle very small volumes of samples and reagents to perform biological assays. Recently, several microwell approaches have been developed that are built on the principle of small volume containment, in which single cells are isolated in nanoliter or sub-nanoliter volumes.[9–11] As a result, secreted molecules are concentrated into a detectable range and secretion from single cells can be quantified. The microwell technique described here, referred to as the single-cell barcode chip (SCBC), is a self-contained chip that can be easily maintained in a conventional CO_2 incubator.[8] It does not require a sophisticated microfluidic control system to operate and thus can be adopted for broad use by researchers and clinicians with minimal engineering background. Using this assay, we have recently demonstrated the importance of paracrine (cell-to-neighbor cell) communication in coordinating the LPS-TLR4 response in macrophage populations.[12]

2. Materials

2.1 *Reagents*

(1) Paired ELISA capture antibody and biotinylated detection antibody for each protein target (up to 16 protein targets possible per chip).
(2) Polydimethylsiloxane (PDMS) supplied as prepolymer and curing agent (Momentive RTV615).
(3) PBS.
(4) PBS + 3% Bovine serum albumin (PBS + 3%BSA) and/or PBS + 5% fetal bovine serum (PBS + 5% FBS).
(5) Allophycocyanin (APC)-Streptavidin.
(6) 70% Ethanol.
(7) Isopropyl alcohol
(8) Cell culture reagents (e.g. medium, serum, antibiotics).

2.2 *Microfluidic supplies*

(1) Master mold for casting flow-patterning chip (Note 1).
(2) Master mold for casting microchamber chip (Note 2).
(3) Poly-l-lysine-coated or epoxy silane-coated glass microscope slides (Erie Scientific).
(4) Glass microscope slide.
(5) Medical microbore tubing (Tygon AAD04103).
(6) Stainless steel tubes (New England Small Tube custom 0.0245/0.0255" OD, 0.5 ± 0.005 Long).
(7) Transparent plates with springs and screws (Note 3).
(8) Nitrogen gas.
(9) Scotch tape
(10) Razor blade
(11) Plastic 100 mm petri dishes

2.3 *Equipment*

(1) Incubator (37°C, 5% CO_2, humidified).
(2) Vacuum desiccator.
(3) Oven.
(4) Switch-controlled pressure-driven microfluidic station (Note 4).
(5) Manual press (Schmidt).
(6) Sonicator (Branson 1510).
(7) Oxygen plasma cleaner (PlasmaEtch PE-25-JW).
(8) Phase contrast microscope and camera (Nikon).
(9) Microarray scanner (Genepix® 4200A from Molecular Devices).

3. Procedure

3.1 *Preparing the antibody barcode slide*

3.1.1 Fabrication of new PDMS chip used for flow patterning (Fig. 1).

(1) Place the flow patterning master mold in a clean plastic petri dish.
(2) Cast wet PDMS (prepolymer:curing agent = 10:1 (w/w)) onto the flow patterning master mold. Remove air bubbles with a vacuum desiccator for 2 hrs.

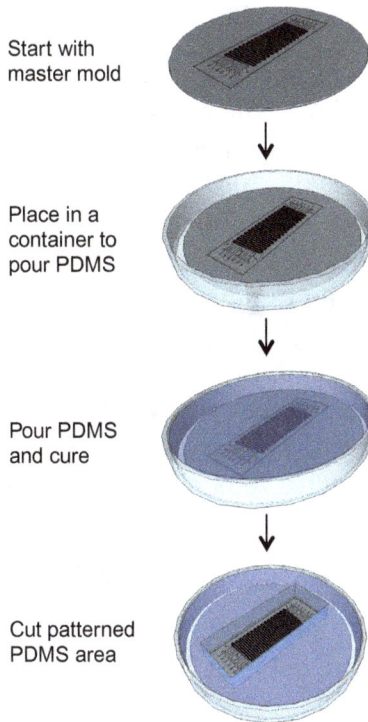

Start with
master mold

Place in a
container to
pour PDMS

Pour PDMS
and cure

Cut patterned
PDMS area

Fig. 1. Fabrication Procedure for the PDMS Flow Patterning and Microchamber Chips.

(3) Heat in an oven for 120 m at 80°C to cure PDMS.
(4) Peel and cut PDMS chip from the mold using a razor blade. Punch holes for inlet and outlet at the marks using manual press.

3.1.2 Clean PDMS flow-patterning chip (if reusing)

(1) Submerge chip in 70% EtOH in 50 mL conical tube. Be sure entire chip is covered and features are in contact with ethanol.
(2) Sonicate for 30 m. Discard ethanol and replace with fresh ethanol. Repeat 1×.
(3) Repeat steps 1 and 2 using isopropyl alchohol in place of ethanol.
(4) Blow dry PDMS with filtered air and dry for 2–3 days. For faster drying, bake the PDMS for 30 m at 80°C (Note 5). Store dry chip in clean plastic 100 mm petri dish.

3.1.3 Bond flow-patterning chip to glass slide

(1) In a clean room or hood, press Scotch tape firmly onto the fea-
tures-side of the PDMS chip and remove. Repeat at least 5 times
to remove dust from the channels.

(2) Gently place the PDMS chip onto a clean poly-L-lysine-coated or
epoxy silane-coated glass slide. Under a microscope, verify that
each channel is free of bubbles and dust. If there are bubbles or
dust particles, do not use those channels for the assay.

(3) Mark the back of the glass slide with the position of the flow- patterning chip. This will indicate the side of the glass slide that does
not contain the antibodies and will help to correctly align the
barcode slide onto the microchamber array (Section 3.3.1, step 4).

(4) Bake for 120 m at 80°C to bond the slide to the chip.

3.1.4 Flow patterning the antibody barcode array

(1) Aliquot capture antibodies into fresh microfuge tubes.
Each barcode array requires 3–4 μL of capture antibody (100–
200 μg/mL).

(2) Cut small segments (~5") of microbore tubing and place stainless steel pin in one end of tube (Fig. 2a and Note 6). Connect
these tubes to the flow station tubes connected to Luer locks via
another pin connection.

(3) Insert Luer lock onto pipet (0.2–2 μL) using an adaptor to create a seal. Place clean pin into antibody solution and pipet 2 μL
into the tubing. Then insert the pin into the PDMS channel
(Fig. 2b). Remove the Luer lock from the pipet and attach to the
flow line (Fig. 2c).

(4) Repeat steps 1–3 for each antibody to be tested using one antibody
per channel. Record the channel number for each antibody.

(5) Turn on flow from nitrogen gas tank and set pressure to 2–3 PSI.
Open flow channels.

(6) Use microscope to check if antibody is flowing. A slight change
in color is observed when the antibody is flowing correctly in the
channel. Continue the nitrogen flow until dry (several hours to
overnight). The color change will disappear when the flow patterning is complete.

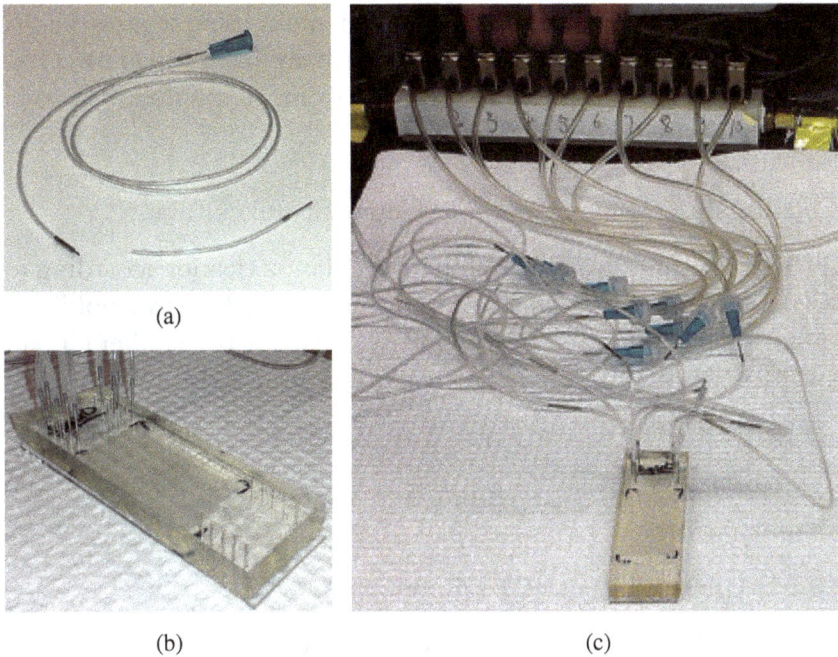

Fig. 2. Assembly of the Device for Microfluidic Flow Patterning of Capture Antiboidies. (a) Clean stainless steel pins are insered into clean Microbore tubing and then this section of tubing is connected to station tubing by another pin. (b) After loading with antibody, clean end of pin is inserted into the PDMS flow Patterning chip. (c) Tubing is connected to the manifolds via Luer locks.

(7) Disassemble device. Submerge barcode slide in PBS+3% BSA or PBS+5% FBS in a 50 mL conical tube and block at RT for at least 120 m or up to 6 hrs.

3.2 *Preparing the microchamber array*

3.2.1 Fabrication of PDMS microchamber array (Fig. 1).

(1) Place the microchamber master mold in a clean plastic petri dish.
(2) Cast wet PDMS (prepolymer:curing agent = 10:1 (w/w)) onto the microchamber master mold. Remove air bubbles with a vacuum desiccator for 2 hrs.
(3) Heat in an oven for 120 m at 80°C to cure PDMS.

(4) Peel and cut PDMS from the master mold using a razor blade. Store in covered petri dish at RT until ready to use (up to 2 weeks).

(5) Autoclave PDMS in clean glass petri dish to sterilize and completely cross-link PDMS.

3.2.2 Condition the microchamber chip for cell culture

(1) Treat the microchamber chip in the plasma cleaner according to manufacturer's instructions to render the surface hydrophilic.

(2) In a 50 mL conical tube, submerge PDMS microchamber chip in PBS+3% BSA or PBS+5% FBS and block for at least 120 m up to 4 hrs at RT.

3.3 *Cell Loading and stimulation (use sterile conditions)*

3.3.1 Seeding adherent cells onto chip (Note 7)

(1) In a biological safety cabinet (BSC) or other sterile environment, prepare single-cell suspension at density of 250,000 cells/mL.

(2) Place paper towels in the BSC. Wash microchamber chip and glass slides 3x with cell media using paper towels to collect liquid. Place microchamber chip on top of glass slide (or other hard clean surface) with wells facing up to so that the chip does not stick to the paper towels.

(3) Seed 150–200 μL of cell suspension onto microchamber array and spread with pipet tip. Allow cells to settle for 1–3 m (should be optimized for cell type being used).

(4) Slowly place second glass slide over microchamber array at an angle to avoid generating flow and displacing the cells. Place device in a 100 mm plastic tissue culture dish.

(5) Check cell density on microscope. Approximately 20% of the wells should contain a single cell. If density is too high or low, remove cells by washing and repeat.

(6) Add media to the bottom of the tissue culture dish and place in CO_2 incubator. Allow cells to adhere (~2 hrs). If not performing assay immediately, add media until it covers the glass slide to prevent the wells from drying out.

3.3.2 Stimulating cells

(1) Prepare stimulation media and antibody barcode slide.

(2) Remove both glass slides and media from the microchamber chip. Wash the chip gently with media 3–5X. Leave media in chip for 1 m each time. Also wash antibody barcode slide 2x with media alone and then 1x with stimulation media.

(3) Place microchamber chip on bottom transparent sealing plate (with screw holes). Place screws in the top transparent sealing plate and set aside until step 6.

If directly stimulating cells in suspension on the chip, go to Section 3.3.1, steps 1–3, then return to step 5.

(4) Pipet 1 mL of stimulation media onto the microchamber chip.

(5) Place the antibody barcode slide onto the chip. Use the marks on the back of the slide to align the barcodes perpendicular to the wells. (*Note: If suspension contains cells, slowly place the glass slide over microchamber array at an angle to avoid generating flow displacing the cells*).

(6) Put 1 mL of stimulation media on top of glass slide and then place top sealing plate. Tighten screws on sealing plates with a screwdriver until firm but avoid cracking the slide (Fig. 3). Place device in a 100 mm plastic tissue culture dish.

(7) Add sterile PBS to tissue culture dish to maintain moisture and prevent evaporation.

(8) Incubate at 37°C in CO_2 incubator until ready to image.

3.3.3 Imaging microchamber array (note 8)

(1) Take a phase-contrast/dark-field/oblique image of the chip at 4x magnification according to microscope specifications (Note 9). Scan the microchamber to record cell number and position per well.

(2) Return to CO_2 incubator until incubation is complete.

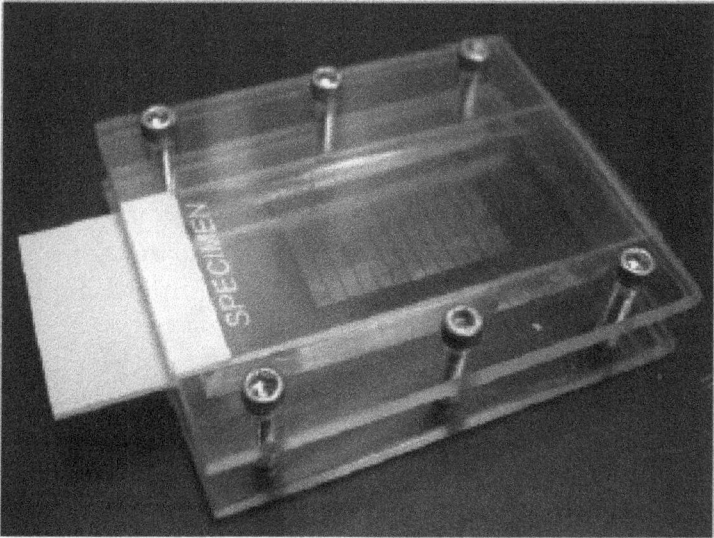

Fig. 3.   Picture of a Fully Assembled Single-Cell Barcode Chip. A high-density antibody array glass slide and a microchamber PDMS slab were clamped together with two transparent plates using springs and screws.

3.4 *Immunoassay*

3.4.1 Device disassembly

(1) Remove device from incubator and place in a clean plastic container (such as a pipet tip box) with PBS+3% BSA. Open the device inside the BSA solution with a screwdriver. Avoid contact with the antibody barcode.

(2) Wash the antibody barcode slide with 1 mL PBS+3% BSA using a pipet to place wash solution in border around the antibody barcode and then tip slide to spread liquid over surface. Remove liquid and repeat 5–6x. After wash steps, gently wipe the edges of the slide dry using a Kimwipe.

3.4.2 Developing the antibody barcode array

(1) Prepare 300 μL of biotinylated detection antibody mixture per slide by combining all detection antibodies in a single solution (use manufacturer recommended dilution in PBS+3% BSA).

(2) Add 300 μL of detection antibody mixture onto the antibody barcode and incubate for 1 hr at RT.

(3) Wash the barcode slide with PBS+3%BSA as described in Section 3.4.1, step 2. Repeat 5–6x.

(4) Add streptavidin-APC diluted 1:100 in PBS+3% BSA and incubate at RT for 30 m protected from light.

(5) Wash with PBS+3% BSA. Repeat 5-6x.

(6) Block slide with PBS+3% BSA for 20 m in dark.

(7) Wash slide with PBS. Repeat 2-3x.

(8) Using a conical tube containing PBS, wash slide by submerging into PBS for 2–3 s. Shake of excess PBS and repeat with a fresh tube of PBS.

(9) Using a conical tube containing deionized water, submerge slide for 2–3 s. Shake off excess water and repeat.

(10) Dry slide by blowing filtered air onto the surface.

(11) Image with microarray scanner. (Can store dry in the dark up to 2 days prior to imaging).

3.5 Data processing and analysis

3.5.1 Count cells

(1) Identify wells (manually or automatically).

(2) Count cells (manually or automatically).

(3) Record cell number per well.

3.5.2 Analyze barcode intensities

(1) Quantify average intensity of barcodes with background subtracted using Genepix® software or other method.

(2) Overlay well grid to match cell number with barcode intensity.

3.5.3 Process raw data matrices

(1) A detection threshold (DT) is calculated from the channels without cells (0-cell background) and is defined as mean + 2 × the standard deviation (SD) of the 0-cell data.

(2) Cytokine production is calculated from pixel intensity minus DT.

(3) To compare across experiments, the threshold is subtracted from the single-cell intensities and the values of all intensities below the threshold are set to zero.

4. Notes

1. The master mold for the flow-patterning chip is silicon etched using the deep reactive-ion etching (DRIE) method from a negative resist mask (design file available upon request). Fabrication of the mask and master mold can be outsourced to a number of companies.

2. The master mold for the microchamber array is silicon etched using the DRIE method from a negative resist mask (design file available upon request). There are two versions of the microchamber array master to accommodate different cell sizes: A 3080-well array (35 μm × 35 μm × 1850 μm, width × depth × length, suitable for macrophages) or a 5440-well array (20 μm × 15 μm × 1850 μm, suitable for T cells). Fabrication of the mask and master mold can be outsourced to a number of companies.

3. Transparent plates with screws must be custom fabricated.

4. Flow patterning is a method that uses microfluidic channels to deliver biomolecules onto a substrate.[13] It requires only microliter volumes and provides specific positioning of the biomolecules. It is relatively inexpensive to set up a microfluidics station to enable flow patterning. For general instructions, see: http://2011. igem.org/Team:EPF-Lausanne/Tools/Microfluidics/HowTo2.

5. Baking ages PDMS, which limits the number of times it can be reused.

6. Stainless steel pins can be reused by following the same procedure used for the PDMS flow-patterning chip (see Section 3.1.2).

7. The above protocol is designed to allow cells to adhere to the microchamber chip prior to stimulation. If cells do not need to be adherent, then seeding and stimulation can be done

simultaneously. In this case, proceed directly to Section 3.3.2 and follow alternate protocol.

8. Cells in the device can be imaged immediately after cell loading prior to cell culture or near the end of the incubation period.

9. Method of well and cell detection will determine the type of image that needs to be recorded.

Acknowledgement

This work was supported by the National Institutes of Health Grant no. U01-CA164252 (to R.F. and K.M.J.).

References

1. Iwasaki A, Medzhitov R. (2010) Regulation of adaptive immunity by the innate immune system. *Science* **327**: 291-295.
2. Lacy P, Stow, JL. (2011) Cytokine release from innate immune cells: Association with diverse membrane trafficking pathways. *Blood* **118**: 9–18.
3. Shalek AK *et al.* (2013) Single-cell transcriptomics reveals bimodality in expression and splicing in immune cells. *Nature* **498**:236–240.
4. Shalek AK *et al.* (2014) Single-cell RNA-seq reveals dynamic paracrine control of cellular variation. *Nature* **509**:363-369.
5. Junkin M,Tay S. (2014) Microfluidic single-cell analysis for systems immunology. *Lab Chip* **14**:1246–1260 .
6. Chattopadhyay PK, Gierahn TM, Roederer M *et al.* (2014) Single-cell technologies for monitoring immune systems. *Nat Immunol* **15**:128–135.
7. Freer G, Rindi L. (2013) Intracellular cytokine detection by fluorescence-activated flow cytometry: Basic principles and recent advances. *Methods* **61**: 30–38 .
8. Lu Y *et al.* (2013) High-throughput secretomic analysis of single cells to assess functional cellular heterogeneity. *Anal Chem* **85**: 2548–2556.
9. Ma C *et al.* (2011) A clinical microchip for evaluation of single immune cells reveals high functional heterogeneity in phenotypically similar T cells. *Nat Med* **17**: 738–743.
10. Han Q, Bradshaw EM, Nilsson B *et al.* (2010) Multidimensional analysis of the frequencies and rates of cytokine secretion from single cells by quantitative microengraving. *Lab Chip* **10**:1391–1400.

11. Han Q *et al.* (2012) Polyfunctional responses by human T cells result from sequential release of cytokines. *Proc Natl Acad Sci USA* **109**: 1607–1612.

12. Xue Q *et al.* (2015) Analysis of single-cell cytokine secretion reveals a role for paracrine signaling in coordinating macrophage response to TLR4 stimulation. *Science Signaling* 8(381):ra59.

13. Delamarche E, Bernard A, Schmid H *et al.* (1997). Patterned delivery of immunoglobulins to surfaces using microfluidic networks. *Science* **276**:779–781.

Chapter 5

Analysis of Tissue Microenvironments Using Decellularized Mammalian Tissues

Huanxing Sun, Yangyang Zhu and Erica L. Herzog[*,†]

Department of Internal Medicine, Section of Pulmonary, Critical Care, and Sleep Medicine, Yale University School of Medicine, New Haven, CT, P.O. 208057, 06520, USA
*erica.herzog@yale.edu

1. Introduction

In adult tissues, homeostasis is maintained by a balance of biochemical and biophysical cues in the organ's microenvironment which integrate the combined contributions of the extracellular matrix (ECM), parenchymal cells, immune cells, and secreted soluble mediators.[1] This complex interplay can be disrupted at any level by processes such as injurious exposures and infections. When these insults can be controlled and the tissue has adequate regenerative capacity, the organ can recover sufficiently to continue functioning. In the setting of overwhelming or persistent injury and/or insufficient repair capacity, tissues can develop pathologic responses characterized

---

[†]Corresponding author.

by disrupted architecture, aberrant cellular phenotypes, and impaired function. In organs such as the lung, these processes result in either rarefication (thinning such as that seen in chronic obstructive pulmonary disease) or excessive remodeling (thickening such as that seen in fibrotic disorders[2]).[3] The processes of injury, repair and remodeling are insufficiently understood and are thought to involve structural cells, resident and recruited inflammatory cells, and activated tissue fibroblasts.[4] Appropriate interactions between these cells require an intact extracellular matrix (ECM) with the proper biochemical and biophysical properties.[5] Given the complexity of these interactions, their *ex vivo* study remains challenging and incompletely achieved. Thus the development of models incorporating salient aspects of the lung microenvironment remains a critical unmet need in this area.

Methods adapted from common bioengineering techniques have emerged as promising tools to investigate the interaction between cells and the tissue microenvironment. One strategy gaining widespread implementation in this regard is the use of scaffolds prepared from decellularized organs.[2] We originally developed the method of mammalian lung decellularization as a platform for the generation of organotypic, biomimetic organs that are suitable for implantation and gas exchange in vivo.[6] Recent application of this method to the study of lung remodeling and repair demonstrates that decellularized lung scaffolds contain physical and biochemical cues that direct cell fate.[7–9] Additionally, acellular matrices derived from fibrotic lungs contain an aberrant ECM that directly influences fibroblast phenotypes[10] and can be used to evaluate the mechanisms governing cellular communication in health and disease.[11] An example of our adaptations of this method to study the lung microenvironment in the context of pulmonary fibrosis is outlined below.

2. Representative Results

2.1 *Structure and composition of decellularized mammalian lungs*

Our initial efforts in this area employed a method based on decellularized whole rat lung that was originally developed by our group as a platform for the *ex vivo* generation of organotypic, biomimetic lung tissue.[6] The decellularized rat lung scaffolds were cut into 2 mm slices suitable for tissue culture. Characterization by H&E staining and scanning electron microscopy (SEM) revealed these slices to possess intact alveolar architecture and the complete absence of cells. These data show that the decellularized lung can be processed into smaller slices potentially suitable for cell culture.

Because fibrosis is most widely studied using genetically engineered mice, we further adapted this technique to the study of mouse lung and simplified the lung decellularization process. We started by harvesting lung tissue from adult C57BL/6 mice. As shown in Fig. 1a, following right ventricular perfusion, intact lungs were removed *en bloc,* snap frozen, sliced, and decellularized. Nucleic acids were removed by benzonase and slices were rinsed and sterilized. Analysis of the resulting mouse lung scaffold slices by H&E staining and SEM demonstrated that the cells were removed by the decellularization process, and that the alveolar septal architecture was maintained (Fig. 1b). Evaluation of decellularization using DAPI staining and Fluorometric DNA quantification confirmed that more than 99% of DNA was removed by the decellularization process (Figs. 1c and d). Immunoblotting for major histocompatibility complex II (MHC-II) as well as β-actin confirmed that the decellularized mouse lung scaffolds were depleted of these cellular markers (Fig. 1e). Immunofluorescence and immunohistochemical staining indicated that structural proteins such as fibrillar collagens, and glycoproteins such as Elastin, Laminin and Fibronectin were preserved in the decellularized matrices, and that most sulfated glycosaminoglycans (GAGs) were removed (Fig. 1f). We and others have applied similar approaches to the normal and fibrotic human lung.[6,11,12] These data show that decellularization of mammalian lung slices produces an acellular matrix scaffold retaining many of the gross anatomic, microstructural, and biochemical properties of native lung, but lacking detectable cellular components. A complete listing of the materials required for the successful decellularization of resected lung tissue is shown in Table 1.

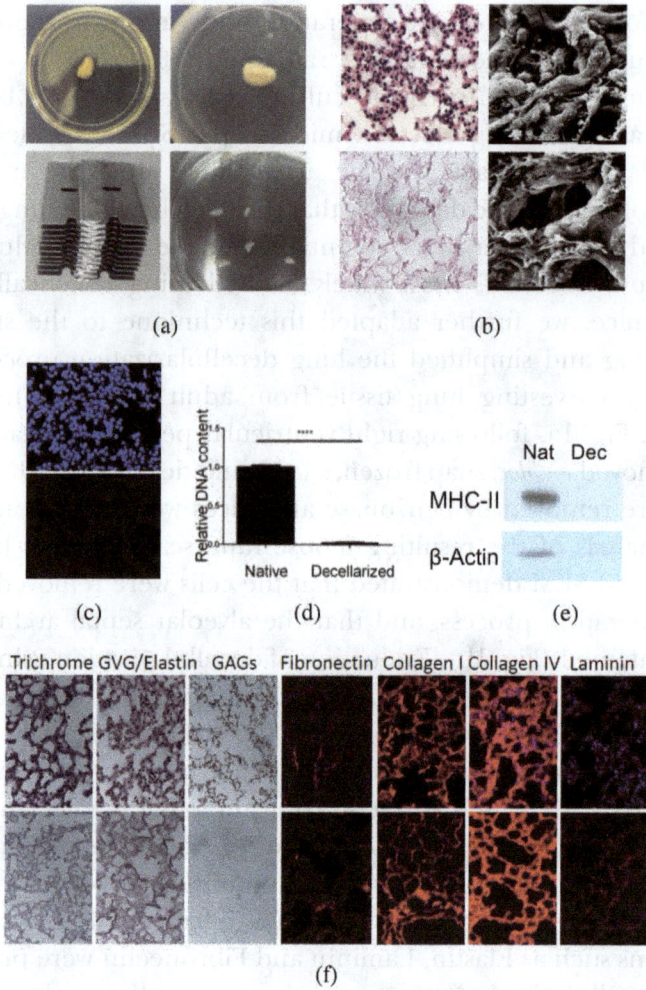

(a)

(b)

(c)   (d)   (e)

(f)

Fig. 1.   (a) Preparation of decellularized murine lung slices. Clockwise from top left: (1) Lavaged and perfused murine lung is placed in a 60 mM Petri dish. which will be transfered at a 100 mM Petri dish and floated on liquid nitrogen to snap freeze the lung. (2) A snap frozen lung lobe. (3) The multi-knife to cut the frozen lung. (4) Frozen lung is cut into slices. (b) H&E staining (left) and SEM (right) of native lung (top) and decellularized lung (bottom) indicates that the appearance of cellular components is absent from the decellularized lung slice. (c, d) DNA content analysis by (c) DAPI nuclear counterstaining and (d) fluorometric quantification of digested scaffold indicates that DNA is absent from the decellularized matrix. (e) Western blot analysis for MHC-II and β-actin reveals that immunogenic and cellular proteins are absent from the decellularized lung.

Fig. 1 (*Figure on facing page*).   (f) Comparison of ECM proteins in native (top) versus decellularized (bottom) murine lung slices. All images are 40× original magnification. Fluorescence images are counterstained with DAPI. Reprinted with permission from American Journal of Physiology — Lung Cellular and Molecular Physiology.[8]

Table 1.   Table of Required Materials.

Name of the reagent	Company	Catalogue number	Comments (optional)
Euthanasia			
Heparin	Sigma-Aldrich	H4784	
SNP	Fluka	71778	
Urethane	Sigma-Aldrich	U2500-500G	180mg/mL in 1XPBS
Deceit Solution			
CHAPS	Sigma-Aldrich	C3023	
NaCI	J. T. Baker	3624-01	
EDTA	Amerixan Bioanalytical	AB00502-01000	
10XPBS	Amerixan Bioanalytical	AB11072-04000	
Benzonase Treatment			
Tris-HCI	Amerixan Bioanalytical	AB14043-01000	pH8.0
BSA	Sigma-Aldrich	A9647-100G	
$MgCl_2$	J. T. Baker	2444-01	
Benzonase Nuclease	Sigma-Aldrich	E1014	Endonuclease used to remove remnant DNA from matrix
Rinsing Solution			
Penicillin/ Streptomycin	Gibco	15140-122	
Gentamicin	Sigma-Aldrich	G-1397-10mL	
Anti-anti	Gibco	15240-062	2%, optional

2.2 *Decellularized lung scaffolds can be used to study macrophage:fibroblast crosstalk*

These scaffolds were used in co-culture experiments to determine whether discrete macrophage populations differentially regulate fibrosis functions that are associated with fibroproliferation. Our previous work had indicated that macrophages expressing the mannose receptor (CD206) might influence fibroblast function.[13] In order to study this question in a simulated lung microenvironment, we performed the following experiments. Flow cytometry was used to isolate M2-like macrophages from the fibrotic mouse lung based on the presence or absence of surface CD206. To determine whether differences exist in the ability of these populations to influence fibroblast, here, decellularized scaffolds prepared from adult mouse lungs were placed in Transwell plates, seeded with a murine fibroblast cell line, and grown in the presence of either total macrophages, CD206^{hi} macrophages, CD206^{lo} macrophages, or no macrophages (Fig. 2a). Our results show that macrophages obtained from the fibrotic mouse lung stimulate fibroblast accumulation (Fig. 2b) with robust stimulation of proliferation and survival (Fig. 2c) but not transformation into myofibroblasts (Fig. 2c) and that these results are largely related to the presence of CD206^{hi} macrophages (Figs. 2b and c). Similar results were seen in decellularized human lungs, where fibroblasts co-cultured with CD206^{hi} monocytes obtained from human volunteers grew more robustly than those grown with CD206^{lo} monocytes or no cells (Fig. 2d).[11]

3. Conclusion

The data presented above support the conclusion that decellularized lung can be used as a novel platform to study various aspects of the tissue microenvironment. Similar approaches exist for most other solid organs[2] which can support additional applications of this modeling system in the context of injury, remodeling, and repair. While by no means a comprehensive recapitulation of the local milieu, the method described above allows investigation of cellular responses in a simulated three dimensional environment of the adult lung.

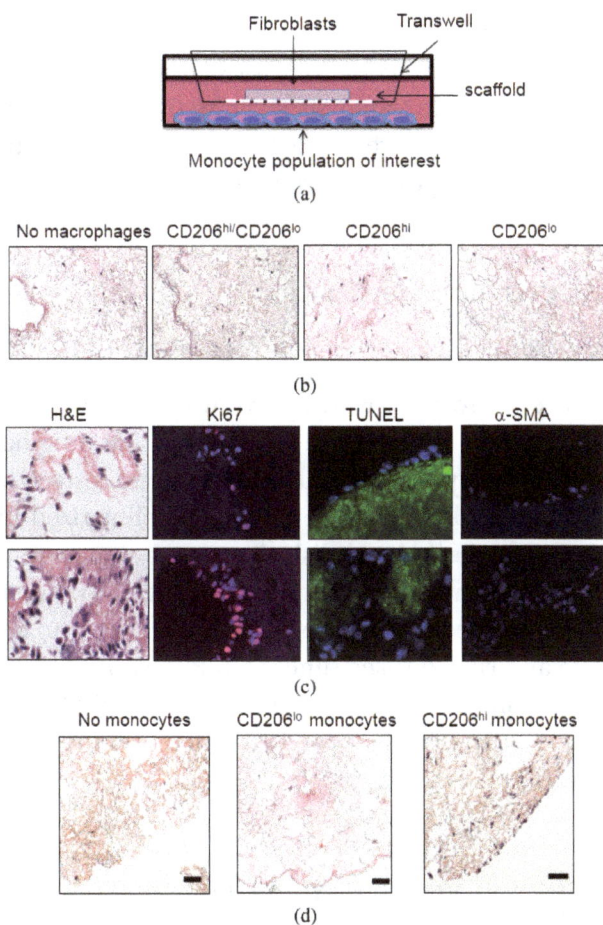

Fig. 2.   (a) Schematic of 3D Transwell culture system. Monocyte-derived cell population resides on the plate's bottom, decellularized lung scaffold slice is placed on the Transwell insert, and fibroblasts are seeded onto the scaffold. (b) Low power H&E images of fibroblast-seeded mouse lung scaffolds grown in the presence of (left to right) no macrophages, CD206$^{hi/lo}$ macrophages, CD206^{hi} macrophages, and CD206^{lo} macrophages obtained from fibrotic mouse lungs. (c) High power 40× view of fibroblast-seeded lung scaffolds grown in the presence of (top) CD206^{lo} macrophages and (bottom) CD206^{hi}- macrophages. From left to right: H&E staining, Ki67 immunostaining, TUNEL detection, and α-smooth muscle actin (α-SMA) immunodetection. (d) Low power images of fibroblast seeded human lung scaffolds slices grown in the presence of (left to right) no primary monocytes, CD206^{hi} primary human monocytes, and CD206^{lo} primary human monocytes. Reprinted with permission from Science Translational Medicine.[11]

As tissue engineering evolves, incorporation of physiologic aspects such as flow, stretch, and mechanotransduction can be incorporated, along with more complex analysis such as transcriptional profiling and live cell imaging, to more fully harness this emerging technology for the *ev vivo* modeling of tissue homeostasis.

4. Detailed Technical Information

4.1 *Preparation of rat lung scaffold slices*

All studies were performed with approval from the Yale University Institutional Animal Care and Use Committee. Rat lung harvest and decellularization was performed as has been extensively described elsewhere.[14,15] To facilitate precise slicing, decellularized lungs were placed in a 60 mm sterile Petri dish that was then positioned within a 100mm Petri dish and floated on liquid nitrogen to snap freeze the scaffolds. Using sterile technique, scaffolds were cut into 2 mm slices by multi-knife and rinsed three times with PBS containing 10% penicillin/streptomycin (Gibco, Grand Island, NY) and 2% Gentamicin (Sigma-Aldrich, St Louis, MO) for 1 hr. Slices were stored in sterile PBS at 4°C until the time of culture.

4.2 *Preparation of mouse lung scaffold slices*

6 to 12 week old mice were treated with a lethal i.p. injection of Urethane (Sigma-Aldrich, St Louis, MO) and heparin (250U/kg) (Sigma-Aldrich, St Louis, MO). Bronchoalveolar lavage and median thoracotomy were performed and the pulmonary vasculature was perfused via right ventricular puncture with PBS. The lungs were removed *en bloc* as we have previously described.[16] To allow precise slicing, each lobe was placed in a 60 mm sterile Petri dish within a 100mm Petri dish and floated on liquid nitrogen to snap freeze the lobes. Using sterile technique, lobes were cut into approximately 1 mm slices by multi-knife and rinsed extensively with PBS containing sodium nitroprusside (SNP) at $1\mu g/mL$ for 1 hr. The slices were transferred to a 15 mL tube and decellularized by 5 mL decellularization solution

(8 mM CHAPS (Sigma-Aldrich, St Louis, MO), 1M NaCl, and 25 mM EDTA in PBS) per lung. The tube was rolled overnight at 37°C. The scaffold slices were then extensively rinsed with PBS, incubated in benzonase nuclease buffer (Benzonase nuclease buffer: 50 mM Tris-HCl, 0.1 mg/mL BSA, 1mM $MgCl_2$, PH 8.0) at room temperature for 10 m, treated with benzonase nuclease in its buffer (Sigma-Aldrich, St Louis, MO) (90U/mL) at 37°C for 1 hr, and then rinsed twice in PBS containing 10% penicillin/streptomycin and 2% Gentamicin. Prepared scaffold slices were stored in sterile PBS at 4°C until the time of culture.

4.3 *Mouse fibroblast coculture with mouse macrophages*

Immediately prior to culture, mouse lung scaffold slices were rinsed with PBS for 5 m, placed in a Transwell insert (Corning Life Sciences), extended with forceps, and incubated at 37°C for at least 10 m to allow adherence to the bottom of the plate. A9 mouse fibroblasts (1×10^4) in 200 μL medium (RPMI + 5% FBS) were drizzled over the lung slice. The seeded slices were placed in the incubator for 30–60 m and then 300 μL of medium was added to the Transwell insert. The bottom of the Transwell plate was incubated with FACs-sorted macrophages of interest (1×10^6 of CD206^{hi}, CD206^{lo}, or CD206^{hi} and CD206^{lo}) in 1.5 mL medium (RPMI + 5% FBS). The medium was changed every 48 hrs. After 7 days, the seeded scaffolds were collected from the upper level of the transwell plate and fixed in 4% PFA for 2 hrs. Samples were then paraffin-embedded and processed for additional analysis.

References

1. Reilkoff RA, Bucala R, Herzog EL. (2011) Fibrocytes: Emerging effector cells in chronic inflammation. *Nat Rev Immunol* 11: 427–435.
2. Song JJ, Ott HC. (2011) Organ engineering based on decellularized matrix scaffolds. *Trends Mol Med* 17: 424–432.
3. Beers MF, Morrisey EE. (2011) The three R's of lung health and disease: Repair, remodeling, and regeneration. *J Clin Invest* 121: 2065–2073.

4. Moore MW, Herzog EL. (2013) Regulation and Relevance of Myofibroblast Responses in Idiopathic Pulmonary Fibrosis. *Curr Pathobiol Rep* **1**: 199–208.
5. Nelson CM, Bissell MJ. (2006) Of extracellular matrix, scaffolds, and signaling: Tissue architecture regulates development, homeostasis, and cancer. *Ann Rev Cell Dev Biol* **22**: 287–309.
6. Petersen TH, Calle EA, Zhao L *et al.* (2010) Tissue-engineered lungs for *in vivo* implantation. *Science* **329**: 538–541.
7. Thomas H, Petersen EAC, Liping Z. (2010) Tissue-engineered lungs for *in vivo* implantation. *Science* **329**: 4.
8. Sun H, Calle E, Chen X *et al.* (2014) Fibroblast engraftment in the decellularized mouse lung occurs via a β1-integrin-dependent, FAK-dependent pathway that is mediated by ERK and opposed by AKT. *Am J Physiol-Lung Cell Mol Physiol* **306**: L463–L475.
9. Calle E, Mendez J, Mahboobe G *et al.* (2015) Fate of distal lung epithelium cultured in a decellularized lung extracellular matrix. *Bioengineering in press.*
10. Parker MW, Rossi D, Peterson M *et al.* (2014) Fibrotic extracellular matrix activates a profibrotic positive feedback loop. *J Clin Invest* **124**: 1622–1635.
11. Zhou Y, Peng H, Sun H *et al.* (2014) Chitinase 3-like 1 suppresses injury and promotes fibroproliferative responses in Mammalian lung fibrosis. *Science Transl Med* **6**: 240ra276.
12. Booth AJ, Hadley R, Cornett AM *et al.* (2012) Acellular normal and fibrotic human lung matrices as a culture system for *in vitro* investigation. *Am J Respir Crit Care Med* **186**: 866–876.
13. Murray LA, Chen Q, Kramer MS *et al.* (2011) TGF-beta driven lung fibrosis is macrophage dependent and blocked by Serum amyloid P. *Int J Biochem Cell Biol* **43**: 154–162.
14. Petersen TH, Calle EA, Colehour MB *et al.* (2011) Bioreactor for the long-term culture of lung tissue. *Cell Transplant* **20**: 1117–1126.
15. Calle EA, Petersen TH, Niklason LE. (2012) Procedure for lung engineering. *J Vis Exp* **49**: 2651.
16. Reilkoff RA, Peng H, Murray LA. (2013) Semaphorin 7a+ regulatory T cells are associated with progressive idiopathic pulmonary fibrosis and are implicated in transforming growth factor-beta1-induced pulmonary fibrosis. *Am J Respir Crit Care Med* **187**: 180–188.

Chapter 6

Defining Innate Immune Pathways with Targeted RNAi Silencing

Feng Qian*

State Key Laboratory of Genetic Engineering and
Ministry of Education Key Laboratory of Contemporary
Anthropology, School of Life Sciences, Fudan University
Shanghai 200438, P. R. China
fengqian@fudan.edu.cn

Summary

To define key components of innate immune signaling requires a robust technique that can easily indentify the loss-of-function phenotype of the gene. RNA interference (RNAi) has emerged as a powerful method to down regulate gene expression and has been widely used in functional gene analysis, target identification and validation. The applications of this technology have identified important regulators of innate immune signal transduction pathways and elucidated cellular functions in the innate immune system.

---

*Corresponding author.

1. Introduction

RNAi is an intracellular mechanism leading to sequence specific post-transcriptional gene silencing.[1,2] The reaction is triggered by endogenous production or artificial introduction of double-stranded (ds) RNA into the cytoplasm of the cell. The dsRNA is rapidly processed by the ribonuclease III-like enzyme Dicer to form small 21–23 nucleotide fragments which are named short interfering RNAs (siRNAs). The siRNA duplexes are then recognized by the RNA interfering silencing complex (RISC), a multi-protein complex with RNase activity, and the siRNA duplexes guide the recognition and degradation of the target mRNA.[3,4]

The innate immune system constitutes the first line of defense against invading microbial pathogens. It plays a crucial role in the early recognition of pathogens, in triggering the inflammatory response, and subsequently in initiating the development of adaptive immune response.[5] RNAi has been extensively used to assess the role of target genes in the function of innate immune cells such as antigen presentation and maturation of dendritic cells,[6,7] phagocytosis and viral recognition by macrophages[8,9] and the activation of NK cells.[10] RNAi has also been used to investigate key players regulating the inflammatory response, e.g. verification of requirement of NF-κB p50 in the production of IL-12 by human dendritic cells[11] and identification of NAIPs as important regulatory proteins of the inflammasomes.[12,13]

Cells of the innate immune system recognize molecular structures that are broadly shared by pathogens, known as pathogen-associated molecular patterns (PAMPs), through a large family of pattern recognition receptors (PRRs).[14] PRR activation initiates signaling transduction pathways that regulate the innate immune response.[15] RNAi has been used to perturb the candidate regulators in signaling pathways and reconstruct the molecular networks underlying innate immune processes.[16,17] With the development of RNAi technology, genome-scale screens have been completed and revealed the cofactors required for TLR signaling,[18] RLR signaling[19] and NOD2 signaling.[20] By incorporating information from diverse molecular profiles, RNAi has proven useful for the study of cellular networks at the host-pathogen interface and has delineated the roles of host factors that interact with pathogens and modulate innate defenses.[21,22]

2. Materials

2.1 *Reagents*

 (1) Lipofectamine 2000 (Invitrogen, Cat. No. 11668-019).
 (2) Opti-MEM I medium (Gibco, Cat. No. 31985-062).
 (3) Ficoll-Paque Plus (GE Healthcare, Cat. No. 17-1440-03).
 (4) RPMI-1640 medium (Gibco, Cat. No. 21875-091).
 (5) Penicillin-Streptomycin (Gibco, Cat. No. 15140-148).
 (6) Trypsin-EDTA (Gibco, Cat. No. 15400-054).
 (7) Human Serum (Lonza, Cat. No. 14-402).
 (8) Amaxa Human Macrophage Nucleofector Kit (Lonza, Cat. No. VPA-1008).
 (9) pLKO.1 (Addgene, Cat. No. 10879).
(10) psPAX2 (Addgene, Cat. No. 12260).
(11) pMD2.G (Addgene, Cat. No. 12259).
(12) AgeI (NEB, Cat. No. R0552S).
(13) EcoRI (NEB, Cat. No. R0101S).
(14) T4 DNA ligase (NEB, Cat. No. M0202S).
(15) QIAquick Gel Extraction Kit (Qiagen, Cat. No. 28704).
(16) Qiagen Plasmid Mini Kit (Qiagen, Cat. No. 12123).
(17) DMEM (Gibco, Cat. No. 11965-092).
(18) Fetal Bovine Serum (Gibco, Cat. No. 16000-044).
(19) Puromycin (Invivogen, Cat. No. ant-pr-1).

2.2 *Equipment*

(1) Incubator (37°C, 5% CO_2, humidified)
(2) Nucleofector Device (Lonza)
(3) Centrifuge

3. Procedure

3.1 *Transient transfection of siRNA*

3.1.1 *Design of siRNA sequence*

General guidelines for selecting siRNA target sequences are based on previous published observation.[23,24]

- Select 21–23 nt sequences in the target mRNA, preferably 50–100 nucleotides downstream from the start codon (ATG).
- Search for sequences 5'- $AA(N_{19})UU$ or 5'– $NA(N_{21})$, or 5'- $NAR(N_{17})YNN$, where N is any nucleotide, R is purine (A, G) and Y is pyrimidine (C, U).
- The GC content should be between 30–52% for proper siRNA functionality.
- Avoid stretches of 4 or more nucleotide repeats.
- Perform NCBI's BLAST search to avoid off-target effects on other genes or sequences.

3.1.2 *Transfection of siRNA by lipofectamine 2000*

(1) Plate target cells (1×10^5/well in 24-well plate) and incubate at 37°C in a CO_2 incubator overnight.
(2) Prepare siRNA-Lipofectamine 2000 complexes. For each sample (1×10^5 cells/well), dilute siRNA (10–50 pmol) in 50 μL of serum free Opti-MEM I medium. Dilute 1–2 μL Lipofectamine 2000 in 50 μL of Opti-MEM I medium and incubate for 5 m at room temperature. Combine the diluted siRNA with the diluted lipofetamine 2000 and incubate for 20 m at room temperature.
(3) Add the siRNA-Lipofectamine 2000 complexes to each well containing cells and 500 μL medium, and incubate for 24–72 hrs in a CO_2 incubator at 37°C.
(4) Evaluate the knockdown efficiency by qPCR or Western comparing to cells treated with a scrambled sequence siRNA for negative control.

3.2 *Nucleofection of siRNA*

Nucleofection is novel transfection method based on a combination of electroporation and cell-type specific reagents, and especially designed for the needs of primary cells and cell lines that are difficult-to-transfect. This protocol describes an example for nucleofection of siRNA in human macrophages which was used to examine viral pathogenesis.[9]

3.2.1 Preparation of Peripheral Blood Mononuclear Cells (PBMCs)

(1) The heparinized peripheral blood or leukocyte-enriched buffy coat should be diluted with 2–4 the volumes of PBS.

(2) Carefully layer 35 mL of diluted blood over 15 mL of Ficoll-Paque Plus in a 50 mL tube.

(3) Centrifuge at $800 \times g$ for 20 m at room temperature in a swinging-bucket rotor with brake off.

(4) PBMCs will form a white band in the middle of the tube. Aspirate the upper layer leaving the mononuclear cell layer undisturbed at the interphase. Collect the mononuclear cell layer into a new 50 mL tube containing 5–7 mL RPMI-1640.

(5) Add RPMI-1640 to 30 mL mark, pellet cells at $300 \times g$ for 10 m at room temperature and discard supernatant. Repeat step 5.

(6) Resuspend PBMC pellet in culture medium and count the cells.

3.2.2 Differentiation of macrophages

(1) Plate $2–3 \times 10^7$ PBMCs per 10 cm dish in 10 mL RPMI-1640 supplemented with 20% human serum, 100 u/mL penicillin and 100 μg/mL streptomycin.

(2) Enrich monocyte by plastic adherence for 2 hrs in an incubator at 37°C, 5% CO_2. Remove the non-adherent cells by $3 \times$ wash with PBS.

(3) Differentiate adherent monocytes for 6 days into macrophages with RPMI-1640 medium with 20% human serum. Change medium on day 3.

(4) Wash adherent macrophages $3 \times$ with PBS on day 6.

(5) Add 2 mL Trypsin/EDTA solution (0.5 mg/mL trypsin and 0.2 mg/mL EDTA in PBS) per dish and incubate for 10–15 m at 37°C.

(6) Add culture medium to stop trypsinization, collect and count the cells.

3.2.3 Nucleofection

(1) Resuspend the cell (5–7×10^5) in 100 μL Nucleofector Solution per sample. Add siRNA (3–30 pmol/sample) in 100 μL of cell suspension. Transfer cell/siRNA suspension into cuvette.

(2) Insert the cuvette into Nucleofector Cuvette Holder and apply the appropriate Nucleofector Program.

(3) Add 500 μL of culture medium to the cuvette and transfer the sample into the 12-well plate. Incubate 24 hrs for cell recovery before changing medium.

(4) Evaluate the knockdown efficiency by qPCR or Western comparing to cells treated with a scrambled sequence siRNA for negative control.

3.3 Stable RNAi with shRNA lentivectors

Short hairpin RNA (shRNA) is also used to silence target gene expression through RNAi. It can be transcribed in cells from a vector containing a Pol III promoter. The choice to use chemically synthesized siRNAs or vector-mediated shRNA depends on several factors such as cell type and time demands. Using siRNA, transient knockdown cells can be obtained with a straightforward and rapid procedure. However, the efficiency of transfection is a major issue for siRNA. Primary and non-adherent cells cannot be transfected with siRNA at high levels. Another technical issue for siRNA is its high degradation in cells. Using siRNA, it is difficult to generate a long-term cell line with efficient knockdown of the target gene. For cells that are difficult to transfect, lentiviral-based shRNA technology remains a successful method for delivery of RNAi. Further, delivery of shRNA into target cells through use of a lentiviral vector system allows for stable integration of shRNA and long-term downregulation of specific target genes. However, the creation of a stable shRNA cell line is time-consuming. This protocol describes an example for generating shRNA constructs, and the packaging and delivery of shRNA lentiviral particles, which was adapted from Addgene's pLKO.1 Protocol (http://www.addgene.org/tools/protocols/plko/).

3.3.1 *Generating shRNA constructs*

(1) Design oligonucleotides to generate the hairpin insert. Choose shRNA target sites as described above (3.1.1). To generate oligonucleotides for cloning into shRNA vector pLKO.1[25], extra nucleotides should be added to generate EcoRI and AgeI restriction enzyme sites and RNA pol III Terminator in forward and reverse chains of oligonucleotides (see below). The shRNA will form a hairpin structure with a CTCGAG loop.

Forward oligo:

5′ CCGG — sense sequences — CTCGAG — antisense sequences — TTTTTG 3′

Reverse oligo:

5′ AATTCAAAAA — sense sequences — CTCGAG — antisense sequences 3′

(2) Perform the oligonucleotide annealing. The forward and reverse chains of oligonucleotides are annealed to form a dsDNA fragment with EcoRI and AgeI sites. The annealing mixture: 5 μL Forward oligo (20 μM), 5 μL Reverse oligo (20 μM), 5 μL 10 × NEB buffer 2, and adjust to 50 μL final volume with H_2O. Heat the mixture to 95°C for 5 m and then let it cool slowly to room temperature (25°C).

(3) Digest the shRNA vector with EcoRI and AgeI. The restriction digestion reaction mixture: 2 μg of shRNA vector, 5 μL 10 × NEB buffer 2, 1 μL AgeI, 1 μL EcoRI, and adjust to 50 μL final volume with H_2O. Incubate at 37°C for 2 hrs. Run the digested products on agarose gel. Use a Qiagen QIAquick Gel Extraction Kit to recover the digested plasmid from the gel.

(4) Ligate the hairpin insert into shRNA vector. Reaction mixture: 20 ng of digested shRNA vector, 2 μL annealed oligo, 2 μL 10 × NEB T4 DNA ligase buffer, 1 μL NEB T4 DNA ligase, and adjust to 20 μL final volume with H_2O. Incubate at 16°C for 4–20 hrs.

(5) Transform and screen for correct shRNA inserts. Use 2 μL of ligation mix to transform competent *E.coli* DH5α. Plate on LB agar plates with 100 μg/mL ampicillin and incubate at 37°C overnight. Innoculate 3–5 colonies into LB broth with 100 μg/

mL ampicillin and grow overnight at 37°C Extract plasmid by using a Qiagen Plasmid Mini Kit. Sequence the plasmids with pLKO.1 sequencing primer (5′ CAAGGCTGTTAGAGAGATAA TTGGA 3′).

3.3.2 *Production of lentiviral particles*

(1) Plate HEK-293T cells at a density of $3-5 \times 10^5$ cells per well of a 6-well plate in DMEM plus 10% FBS without antibiotics. Incubate cells at 37°C in a CO_2 incubator overnight.

(2) For each transfection sample, prepare plasmid mixture: $1\mu g$ pLKO.1 shRNA plasmid, 750 ng packaging plasmid (e.g. psPAX2), 250 ng envelope plasmid (e.g. pMD2.G) in 250 μL Opti-MEM I medium. Dilute 10 μL Lipofectamine 2000 in 250 μL of Opti-MEM I medium and incubate for 5 m at room temperature.

(3) Combine the plasmid mixture with diluted Lipofetamine 2000 and incubate for 20 m at room temperature. Add the plasmid-Lipofectamine 2000 mixture to each well containing cells and 2 mL medium, and change medium after 12–15 hrs.

(4) Incubate cells for additional 24–48 hrs in a CO_2 incubator at 37°C. Harvest and filter the lentiviral supernatants through a 0.45 μm filter to remove cellular debris.

3.3.3 *Target cell infection and selection*

(1) Plate target cells and incubate at 37°C in a CO_2 incubator overnight.

(2) Change to fresh culture medium and add lentiviral particle solution (0.1–1mL). Incubate cells at 37°C overnight.

(3) Change to fresh medium after 24 hrs and select the infected cells by adding puromycin (1 to10 $\mu g/mL$).

(4) Evaluate the knockdown efficiency by qPCR or Western comparing to cells treated with scrambled shRNA for negative control.

4. Notes

(1) Several computer programs have been developed for selecting siRNA and shRNA target sequences, such as: siRNA at WHITE HEAD (http://sirna.wi.mit.edu), Invitrogen Block-iT RNAi Designer (http://rnaidesigner.lifetechnologies.com), Dharmacon siDESIGN Center (http://dharmacon.gelifesciences.com/design-center), and The TRC shRNA Design (http://www.broad institute.org/rnai/public/seq/search).

(2) Multiple target sequences for each gene should be selected to increase the probability of obtaining knockdown. Using two different siRNA or shRNA that target the same gene can also avoid concerns about off-target effects.

(3) Experimental controls can include the following samples: Untreated cells, mock-transfected or mock-infected sample, negative control (non target or scrambled target), positive control (targeting an endogenous or reporter gene).

(4) Knockdown efficiency can be measured at the mRNA level by RT-qPCR and protein level after 48 hrs.

(5) Lentiviral particle experiments should be carried out under Biological Safety Level 2 (BL2) or higher.

(6) For puromycin selection, if the concentration of puromycin for the target cell type is unknown, a titration test is necessary.

Acknowledgement

This work was supported in part by the National Natural Science Foundation of China (Grant No. 81370464) and Doctoral Fund of Ministry of Education of China (Grant No. 20130071120031).

References

1. Bosher JM, Labouesse M. (2000) RNA interference: Genetic wand and genetic watchdog. *Nat Cell Biol* **2**: E31–E36.

2. Dykxhoorn DM, Novina CD, Sharp PA. (2003) Killing the messenger: Short RNAs that silence gene expression. *Nat Rev Mol Cell Biol* **4**: 457–467.

3. Elbashir SM, Harborth J, Lendeckel W *et al.* (2001) Duplexes of 21-nucleotide RNAs mediate RNA interference in cultured mammalian cells. *Nature* **411**: 494–498.
4. Chiu YL, Rana TM. (2002) RNAi in human cells: Basic structural and functional features of small interfering RNA. *Mol Cell* **10**: 549–561.
5. Janeway CA, Jr., Medzhitov R. (2002) Innate immune recognition. *Annu Rev Immunol* **20**: 197–216.
6. Hill JA, Ichim TE, Kusznieruk KP *et al.* (2003) Immune modulation by silencing IL-12 production in dendritic cells using small interfering RNA. *J Immunol* **171**: 691–696.
7. Moita CF, Chora A, Hacohen N *et al.* (2012) RNAi screen for kinases and phosphatases that play a role in antigen presentation by dendritic cells. *Eur J Immunol* **42**: 1843–1849.
8. Hatsuzawa K, Tamura T, Hashimoto H *et al.* (2006) Involvement of syntaxin 18, an endoplasmic reticulum (ER)-localized SNARE protein, in ER-mediated phagocytosis. *Mol Biol Cell* **17**: 3964–3977.
9. Kong KF, Delroux K, Wang X *et al.* (2008) Dysregulation of TLR3 impairs the innate immune response to West Nile virus in the elderly. *J Virol* **82**: 7613–7923.
10. Borg C, Jalil A, Laderach D *et al.* (2004) NK cell activation by dendritic cells (DCs) requires the formation of a synapse leading to IL-12 polarization in DCs. *Blood* **104**: 3267–3275.
11. Laderach D, Compagno D, Danos O *et al.* (2003) RNA interference shows critical requirement for NF-kappa B p50 in the production of IL-12 by human dendritic cells. *J Immunol* **171**: 1750–1757.
12. Kofoed EM, Vance RE. (2011) Innate immune recognition of bacterial ligands by NAIPs determines inflammasome specificity. *Nature* **477**: 592–595.
13. Zhao Y, Yang J, Shi J *et al.* (2011) The NLRC4 inflammasome receptors for bacterial flagellin and type III secretion apparatus. *Nature* **477**: 596–600.
14. Janeway CA, Jr. (1989) Approaching the asymptote? Evolution and revolution in immunology. *Cold Spring Harb Symp Quant Biol* **54 Pt 1**: 1–13.
15. Akira S (2009) Pathogen recognition by innate immunity and its signaling. *Proc Jpn Acad Ser B Phys Biol Sci* **85**: 143–156.
16. Amit I, Garber M, Chevrier N *et al.* (2009) Unbiased reconstruction of a mammalian transcriptional network mediating pathogen responses. *Science* **326**: 257–263.

17. Chevrier N, Mertins P, Artyomov MN *et al.* (2011) Systematic discovery of TLR signaling components delineates viral-sensing circuits. *Cell* **147**: 853–867.

18. Chiang CY, Engel A, Opaluch AM *et al.* (2012) Cofactors required for TLR7- and TLR9-dependent innate immune responses. *Cell Host Microbe* **11**: 306–318.

19. Pulloor NK, Nair S, Kostic AD *et al.* (2014) Human genome-wide RNAi screen identifies an essential role for inositol pyrophosphates in Type-I interferon response. *PLoS Pathog* **10**: e1003981.

20. Lipinski S, Grabe N, Jacobs G *et al.* (2012) RNAi screening identifies mediators of NOD2 signaling: implications for spatial specificity of MDP recognition. *Proc Natl Acad Sci U S A* **109**: 21426–21431.

21. Brass AL, Huang IC, Benita Y *et al.* (2009) The IFITM proteins mediate cellular resistance to influenza A H1N1 virus, West Nile virus, and dengue virus. *Cell* **139**: 1243–1254.

22. Krishnan MN, Ng A, Sukumaran B *et al.* (2008) RNA interference screen for human genes associated with West Nile virus infection. *Nature* **455**: 242–245.

23. Reynolds A, Leake D, Boese Q *et al.* (2004) Rational siRNA design for RNA interference. *Nat Biotechnol* **22**: 326–330.

24. Tuschl T. (2006) Selection of siRNA Sequences for Mammalian RNAi. *CSH Protoc*, http://csh protocols.cshlp.org/content/2006/1/pdb.prot4339. Full.

25. Moffat J, Grueneberg DA, Yang X *et al.* (2006) A lentiviral RNAi library for human and mouse genes applied to an arrayed viral high-content screen. *Cell* **124**: 1283–1298.

Chapter 7

ImageStream Methodologies for Flow Cytometry with High Resolution Microscopy

William J. Housley* and Ewa Menet[†,‡]

*Department of Neurology, Yale University School of Medicine, 300 George St. Room 353, New Haven, CT 06511 USA
†Department of Laboratory Medicine, Yale Laboratory Medicine Hospital, TAC S631 P.O. Box 208035 New Haven, CT 06520 USA
‡ewa.menet@yale.edu

1. Introduction

Amnis Imagestream technology is a unique, novel flow cytometry platform that allows for detailed single-cell imaging of fluorescently labeled cells. It represents a merger of confocal imaging, and flow cytometry and offers a new capability in cell imaging that was previously impossible. The analysis software allows for step-by-step analysis of multiple parameters simultaneously without requiring significant previous experience in imaging quantitation programs. While lacking the resolution of standard confocal or the throughput of flow cytometry, the technology has many applications for quantitative

---

‡Corresponding author.

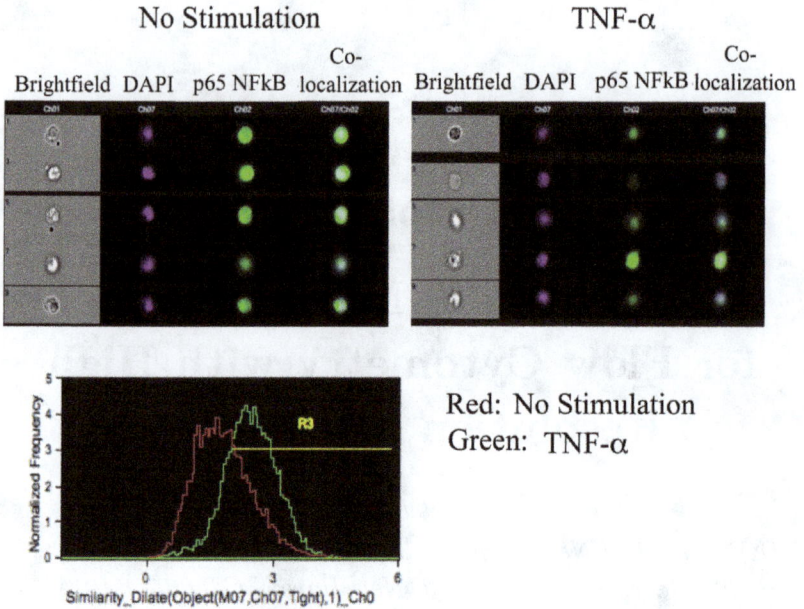

Fig.1.   Example Nuclear Localization of p65 NFκB.

CD4 T cells were stimulated with 50 ng/mL TNF-α for 30 minutes. Nuclear localization was determined by Pearson co-efficient co-localization of DAPI (purple) and p65 NFκB (Green) by Amnis Ideas software. Below, overlay of unstimulated (red) and TNF-α stimulated (Green).

co-localization of fluorescent markers, imaging cell–cell interaction, cell cycle analysis, DNA damage and repair and quantitation of nuclear translocation of transcription factors. A common use for Imagestream is transcription factor nuclear localization after stimulation. The ability to accurately quantitate the amount of a transcription factor that has translocated from the cytoplasm to the nucleus in thousands of cells was previously impossible with confocal microscopy. Figure 1 shows an example of determining nuclear localization of p65 NFκB after TNF-α stimulation in *T* cells and similar protocols have been employed in monocytes and dendritic cells. In addition, Imagestream has been employed to demonstrate close proximity of the receptors for IFNγ and Fcγ in membrane microdomains (1) co-localization of the *B* cell receptor (BCR) with signaling molecules (2) lineage of origin of lung epithelial cells and (3) age-dependent differences in translocation of transcription factors in primary monocytes (4) Individual labs are continually developing new uses for this technology demonstrating its versatility and potential Principles of the technology. Hydrodynamic focusing is used to position single cells within a core stream that are passed through the optics which detects a light microscopic image of each cell as well up to 10 fluorescent markers. Either 6 or 12 images are taken of each cell (depending on the standard or upgraded package for the Mark II). Imagestream employs the time-delay-integration (TDI) detection technology used by CCD cameras which allows measure of up to 10 fluorescent markers with up to 12 images taken of each cell. Amnis Imagestream can only be performed on single-cell suspensions so it is less appropriate for use with some adherent cell populations. Imagestream allows for quantitation of thousands of cells simultaneously rather than a small number of cells within the field of view of the confocal. This represents a major improvement in the sensitivity of the assay to small changes compared to confocal and reduces experiment to experiment variability.

As flow cytometry is a commonly used tool for assessing immune function, we will assume a basic level of flow cytometry experience in this discussion of the Amnis Imagestream^{x}. The protocol for staining and running cells by Amnis Imagestream are similar to those used for standard flow cytometry. Briefly, cells are first labeled with fluorescently tagged antibodies. For nuclear localization experiments, DNA is stained with DAPI or a similar DNA intercalater. While the Amnis is an extremely powerful new tool to analyze single cell function, this chapter will review the current equipment and software for the Amnis imagers, setting up an initial experiment and best practices to overcome challenges we have come across. Finally, we will discuss three papers that have been published with data generated using the Yale Amnis Imagestream^{x} and how the Imagestream allowed analyses that were previously impossible.

2. Equipment Options and Selection

Four Amnis imaging machines exististed in the market at the time of publication: IS100, Imagestream^{x}, Imagestream^{x} Mark II and Flowsight. The first commercially available Amnis instrument was the IS100. Very few of these machines are currently in use as they were quickly replaced by the Imagestream^x. Our primary experience is with the Imagestream^x and this chapter will focus on the use of this machine with discussions of the new fluidics and data capture software for the Mark II instrumentation. The Flowsight is a new, pared down version of the Imagestream^x. However, it is only available with a 20^x magnification, which significantly reduces resolution. In comparison, the Imagestream^x has a reported resolution of $0.1/0.25/1.0$ μM^2 (for $60^x/40^x/20^x$). However, the 20^x optics on the Flowsight only have a 2 μM^2 resolution, making it lower than even the 20^x optics on the Imagestream^x. While we do not have experience with the Flowsight, resolution of the 40^x optics on the Imagestream are often too low for detailed quantitation, suggesting that the Flowsight will be limited in its capacity for detailed imaging. When selecting the appropriate instrument, potential users should consider all possible uses for the machine and choose optics and fluidics appropriate to the planned studies.

Both the Imagestream^x and the Flowsight have optional 96 well autosamplers. Autosamplers need frequent monitoring for alignment of the sample injection head and preventing bubbles and clogs in the Imagestream^x instruments that interfere with automated sampling. The dramatically restructured fluidics on the Mark II instrument are expected to improve these concerns. The Extended Deep Field (EDF) upgrade significantly improves resolution and focuses across the entire cell, which is particularly important if you are using the Imagestream for FISH analysis or identifying smaller cellular structures throughout a cell.

The Amnis Imagestream^x has posed many challenges, particularly with software and fluidics. The Mark II machine has corrected many of these problems and represents a significant improvement over the Imagestream^x. This enhancement includes redesigned fluidics to decrease bubbles and clogging and new software that is more intuitive and user friendly.

3. Software

Amnis' proprietary image analysis software, IDEAS, is designed with a step-by-step walk through to allow users with limited experience in image processing to get started quickly in analyzing data. Imagestream generates large file sizes that require substantial image processing power on any computer used for analysis. The IDEAS software can be run only on Windows™-based operating systems and is not compatible with MAC OS. However, it can be run using a windows environment such as Parallels or VMWare. Amnis offers the IDEAS software for free from their website which may be useful for users with sufficient processing power on their personal computers to analyze image data. The processing of large data sets is lengthy and currently may take days for large data sets. Presently, users can also send captured data to Amnis directly for processing.

For assistance in getting protocols started, Amnis has setup an online knowledge base (https://www.amnis.com/login.html). This is a good resource for accessing protocols currently in use by other labs. While these protocols may still requiring adjustment for individual laboratory uses, our experience has been that they represent a good starting point for successful imaging. This website also supports sending data directly to Amnis to get assistance from software specialists.

4. Getting Started

Compensation and sample staining for the Amnis Imagestream have additional considerations beyond flow cytometers. Here we will discuss some of the best practice advice to develop staining protocols and other factors that should be taken into account when planning an experiment.

4.1 *Selecting dyes and staining*

It is possible to detect over 10 colors on the Imagestream^x, however multicolor imaging requires considerable care. Imagestream generally requires higher antibody concentrations than flow cytometry (2–10 fold higher) and antibody concentrations should be optimized for individual uses. It is also useful to label cells with the lowest antigen abundance using the brightest fluorophores and vice versa. One way that some users have gotten around fluorophores with significant spectral overlap such as A488 and PE is to choose antibodies that

Table 1.   Example of Staining Panel.

	Excitation (nM)	Band (nM)
Brightfield	488	430–480
FITC/A488/GFP	488	480–560
PE	488	560–595
PI	488	595–660
PerCP-Cy5.5	488	660–745
SSC/PE-Cy7*	488	745–800
DAPI/Pacblue	405	430–505
PacOrange	405	505–570
Brightfield	405	570–595
QDOT625	405	595–660
APC/AF647	658	660–745
APC-Cy7	658	745–800

Yellow- starting combination of channels
Blue- secondary color combination
Orange- final color combination for 9 channel analysis
*for imagestreamx Mark II

stain different compartments within the cell (i.e. A488 for the cell surface versus PE for nuclear stains). Finally, you can also get creative with potential cross-talking channels/fluorophores and you can use them as a "dump channels", to dump the subpopulation of cells you do not need within this channel (e.g. lineage dump or viability check). An example of good working combinations is shown in Table 1. Certain tandem dyes may show emission from the donor fluorophore. For example, using APC-Cy7 and APC together may result in overlapping fluorescence that cannot be compensated, making tandem dyes less preferred. In addition, several dyes are excited by more than one laser and thus should not be used in a single panel. For example DRAQ5 is excited by both the 561 nm and the 658 nm lasers and detected on channel 5 or 11 while APC can only be detected on channel 11. Therefore, if you choose DRAQ5 as a fluorophore, you cannot also stain with APC, this further limits the staining options. On the Imagestream[x], Side-Scatter is collected from the 785 nm excitation laser and detected at 745–800 nm. Both PE-Cy7 and PE-AF750 are detected at 745–800 nm, making it impossible to collect SSC and these fluorophores simultaneously. The Imagestream^x Mark II has corrected for this by moving the SSC detection spectrum and should not have the same issue.

When starting a new protocol, we begin with 3–4 antibodies on bright, well-separated channels. Further fluorophores can be optimized first by analysis on flow. For cell types such as monocytes/macrophages and dendritic cells with significant autofluorescence, unstained cells should be run during optimization steps to identify relevant autofluorescence for the population of interest. Best results are obtained when using at least 1 million cells in 50 μL of buffer and filtering the cells prior to running. The Amnis sample preparation guide includes these suggestions.

4.2 Compensation

As in standard flow cytometry, Amnis requires an unstained control and single stained samples to setup compensation. The IDEAS software determines which samples are single stained and calculates compensation automatically. While the optics allow staining on 10 channels simultaneously, dim channels may be difficult to differentiate. Despite high laser intensity on the Imagestream, the low-end detection may show poor resolution of low abundance molecules. Unlike in flow, the compensation is not applied to the cells during data collection. As such, only uncompensated images of the cells are viewed as they are collected.

4.3 Imaging rare cell populations

Imaging rare cell populations requires collection of a sufficient number of cells and results in extremely large file sizes and lengthy data analysis times. The rare cell population must express highly abundant surface markers to make it clearly identifiable from other populations. The flow rate on the Mark II is now 5000 cells/s, which will improve the ability to analyze rare cell populations. Pre-sorting cells by magnetic bead sorting or FACs prior to staining for Amnis may improve collection of desired rare cell populations.

5. Applications

Three examples of recent findings using Amnis Imagestream are discussed below with particular emphasis on the novel findings possible only with Imagestream.

5.1 *Transcription factor nuclear localization*

Qian *et al.* investigated the decline in immune function in elderly patients. In particular, they focused on the increased cytokine expression from monocytes stimulated through TLR5. Peripheral blood mononuclear cells from young or older subjects were stimulated with flagellin, fixed and stained with CD14 PE. NFκB is a key central regulator of immune responses. After TLR5 stimulation, p65 NFκB translocates into the nucleus and drives transcription of many down-stream inflammatory genes. To determine changes in NFκB localization, the cells were then permeabilized and stained for p65 NFκB and DAPI. After gating on CD14+ monocytes, the Amnis IDEAS software was used to measure overlap of p65 NFκB and DAPI stains to determine the amount of nuclear NFκB before and after stimulation. Cells from elderly subjects were found to have increased constitutive nuclear localization of NFκB, but increased nuclear NFκB to a similar level after stimulation with flagellin suggesting a dysregulation of the NFκB pathway in elderly patients. The imagestream technology allowed for the comparison of nuclear localization of NFκB in thousands of CD14+ monocytes that would have been impossible by confocal microscopy alone.[1]

5.2 *Receptor co-localization and intracellular localization*

Bezbradica *et al.* investigated the interaction and localization of IFNγR1 and FcγR1 in murine bone marrow derived dendritic cells (BM DCs) using the Imagestream[x]. Using Alexafluor 488 labeled latex beads and APC labeled IFNγR1, they showed that IFNγRI localized to the phagocytic cup in BM DCs. FcγR1 was preloaded in membrane microdomains consistent with lipid rafts prior to phagocytosis. Both IFNγR1 and FcγR1 were present in these microdomains suggesting close proximity of the receptors. BM DCs were selected over BM macrophages as the macrophages lost their native morphology in suspension. This circumstance highlights one difficulty of these assays in that the cell morphology must be maintained in order to have an accurate measurement of intracellular localization. As discussed previously, the IDEAS analysis software requires significant experience and all analysis of Imagestream data in this publication was performed at Amnis.[2] Using the Amnis Imagestream[x] allowed the analysis of thousands of cells with quantitation of both receptor co-localization and association to phagocytic cups while staining with multiple colors for surface markers and intracellular localization.

5.3 *BCR signaling and cell cycle analysis*

Khalil *et al.* studied the difference in co-localization of the BCR with the signaling molecules SHP1 and SHIP1 between germinal center and non-germinal center murine *B* cells. Germinal center *B* cells exhibited greater constitutive SHP1 and SHIP1 association with the BCR. Signaling through the BCR in non-germinal center *B* cells resulted in dissociation and repolarization of SHP1 and SHIP2 away from the BCR. This dissociation was not evident in germinal center *B* cells. The authors also used the Imagestream to compare SHP-1/ BCR co-localization during the cell cycles G1/S/G2, and M phases. Cell cycle analysis was performed by injection of BrdU into immunized mice. Total splenocytes were stimulated with b-7-6 anti-Mu IgG to stimulate *B* cell proliferation and stained for BrdU and DAPI to determine how SHP1 was associated with the BCR throughout the cell cycle.[3] The Imagestream allowed for simultaneous analysis of co-localization and cell cycle that would not be possible by either flow cytometry or confocal alone.

5.4 *Specificity of cell phenotyping*

Kassmer *et al.* investigated the cellular progenitors of lung-resident epithelial cells. Bone-marrow derived cells of hematopoietic and non-hematopoietic origin were derived and transferred into Surfactant protein C (SPC) knockout mice. 2–6 months after transplant, single-cell suspensions of lung cells were pre-enriched for type 2 pneumo-cytes and Amnis Imagestream used to identify donor-derived SPC+ type 2 alveolar cells. Mice transferred with non-hematopoietic cells contained donor derived SPC+ type 2 alveolar cells, while those transferred with hematopoietic cells did not, suggesting that the non-hematopoietic cells are the source of bone-marrow derived lung epithelial cells. These experiments used Amnis Imagestream in order to positively identify donor derived lung epithelial cells. The Amnis allowed for a sensitivity of detection impossible with standard flow cytometry to identify type 2 pneumocytes and discriminate true stain-ing from autofluorescence. As SPC is usually expressed in lamellar bodies with a vesicular staining pattern, cells were identified by hav-ing at least two bright spots of SPC localization within the cytoplasm. Autofluorescent cells were removed from the gating analysis by elimi-nating cells in which SPC and cytokeratin were co-localized.[4]

6. Conclusion

The Amnis Imagestream imaging technology represents a revolution in our ability to simultaneously image single-cell populations with multiple fluorescent labels at a high resolution. The ability to gate on specific sub-populations from a population of thousands of cells and then assessing co-localization, morphology, cell cycle and other parameters is unprecedented. Here, we have provided some guid-ance to getting initial experiments started and discussed approaches for working with this new technology. We hope to facilitate the suc-cessful adoption of this technology to new techniques and research.

Acknowledgements

We are grateful to the many users of the Yale facility that have con-tributed their advice and experience.

References

1. Qian F *et al.* (2012) Age-associated elevation in TLR5 leads to increased inflammatory responses in the elderly. *Aging cell* **11**:104.
2. Bezbradica JS, Rosenstein RK, DeMarco RA *et al.* (2014) A role for the ITAM signaling module in specifying cytokine-receptor functions. *Nat Immunol* **15**: 333.
3. Khalil AM, Cambier JC, Shlomchik MJ. (2012) B cell receptor signal transduction in the GC is short-circuited by high phosphatase activity. *Science* **336**:1178.
4. Kassmer SH, Bruscia EM , Zhang PX *et al.* (2012) Nonhematopoietic cells are the primary source of bone marrow-derived lung epithelial cells. *Stem cells* **30**:491.

Chapter 8

First Responders: Laboratory Methods to Assess Human Neutrophils

Jose Thekkiniath*, Yi Yao* and Ruth R. Montgomery[†]

Department of Internal Medicine, Yale University School of Medicine, New Haven, CT, P.O. Box 06520, USA
[†] *ruth.montgomery@yale.edu*

1. Introduction

Neutrophils, or polymorphonuclear leukocytes (PMN), are short-lived terminally differentiated leukocytes with powerful anti-microbial functions.[1,2] In humans, PMN are the most abundant leukocytes — generally ~4×10^6/mL in blood and about 70% of the circulating white cells.[3] PMN are produced by the bone marrow at ~10^9 cells/kg/day and are the first cells recruited to the site of infection or injury, where they release potent reactive oxygen and nitrogen intermediates along with granules containing anti-microbial agents.[4] PMNs have been recognized for more than a

---

*These authors contributed equally.
[†] Corresponding author.

century for their phagocytic function and anti-microbial potency, however, recent studies reveal more complex roles for PMN in immune responses including cytokine production, extracellular trap formation, and regulation of adaptive immunity.[1,5–7]

PMN are highly sensitive to experimental manipulation and easily become activated during preparation, thus investigators have developed specialized experimental methods for their study. Activation can be quantified by multiple measures including receptor expression, involvement of signaling pathways, cytokine production, and prolonged apoptosis (Fig. 1). In this chapter, we discuss the experimental techniques typically employed for harvest and assessment of the phenotype and functionality of PMNs, including one-step isolation of PMN by dextran sedimentation[8]; methods for quantifying PMN function by flow cytometry through detection of

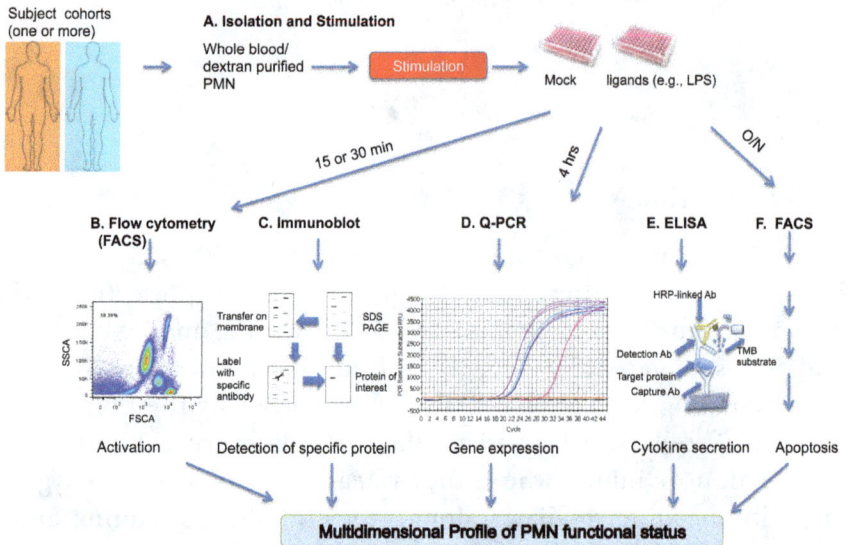

Fig.1. Coordinated Profiling of PMN Function. Schematic illustration of PMN isolation and stimulation (a) and functional assays in response to receptor agonists/ligands. The ligand LPS is used as an example here. Functional assays include (b and f) flow cytometry, (c) Immunoblot, (d) PCR (Q-PCR) and (e) ELISA. Actin is used as reference gene in Q-PCR. O/N indicates overnight incubation for quantitation of apoptosis.

activation markers and cytokine production and activation-dependent delay in apoptosis[9-12]; detection of relevant signaling proteins by immunoblot[12]; gene expression by Q-PCR;[13] and chemokine/cytokine secretion by ELISA.[12] These assays may be used to identify responses to particular agents or functional differences between cohorts of subjects. Furthermore, many of these assays can be assessed in parallel for a detailed profile of PMN function. Specifically, functions of PMN from different cohorts of subjects, such as younger and older, responders and non-responders, or an individual subject before and after treatment, can be assessed first at baseline and also in response to stimulation. Selection of cohorts for comparison and relevant control groups requires care for consistency in sample collection and methods. From the potential measurements (Fig. 1), investigators should select protocols feasible with the cell numbers available and laboratory environment and instrumentation.

Isolation of PMNs (Fig. 1a) is required for some functional assays of PMNs to assess response to stimuli (e.g. LPS) in comparison with unstimulated cells. Flow cytometry, a method that measures several cellular features rapidly, can be used for the quantitative measurement of marker expression and cytokine production in stimulated PMN (Fig. 1b). An immunoblot is used to assess the specific signaling intermediates in PMN pathways that may contribute to cohort-specific changes in immune responses (Fig. 1c). PMN transcriptional response to stimuli is assessed by quantitative PCR (Q-PCR; Fig.1d). ELISA is performed on the supernatants of cultured PMNs to validate whether altered cytokine mRNA levels (e.g. IL-8) are consistent with protein production (Fig. 1e). Apoptotic characteristics of stimulated PMN quantify prolonged lifespan in response to stimulation and can be assessed by flow cytometry (Fig.1f).

2. One-step Isolation of PMNs by Dextran Sedimentation

2.1 *Materials*

(1) Anticoagulant: Preservative-free sodium heparin (1000 U/mL).
(2) 3% (w/v) Dextran solution: dissolve 30 g/L of Dextran (average molecular weight 60,000–90,000) in saline (0.9% NaCl) and filter the solution.

(3) Saline solutions: 0.2% and 1.6% (w/v) NaCl.
(4) Hank's balanced salt solution (HBSS) without Ca^{2+} or Mg^{2+} (sterile, endotoxin free)

2.2 Methods

(1) Mix anti-coagulated blood 1:3 (blood to dextran) with 3% dextran in a 50 mL polypropylene tube at 4°C. Mix gently and thoroughly by repeated inversions (7–10 times) and let sit for 20–35 m at room temperature (RT). PMNs will sediment faster if the tubes are kept at a 45° angle.
(2) Collect the supernatant and wash away dextran by centrifuging at $250 \times g$ for 10 m at RT. Aspirate supernatant and re-suspend pellets in 0.1–0.5 mL HBSS; can combine multiple pellets.
(3) Hypotonic lysis of erythrocytes. With tube on ice, lyse red cells by adding 10 mL 0.2 % saline (4°C) to the cell suspension. Pipette up and down very gently until red cells are lysed (~ 30 s or 5–7 times pipetting). Immediately restore isotonicity by adding 10 mL 1.6% saline. Add HBSS to fill to 50 mL to restore cell integrity. Spin at $250 \times g$ for 10 m at RT.
(4) Aspirate supernatant which contains red cells and red cell ghosts. Re-suspend pellet in 0.5 mL HBSS and count the cells using a hemocytometer.

2.3 Notes

(1) PMNs are short-lived and deteriorate rapidly. For best results, use the cells within 2 hrs of collection of blood.
(2) For PMN isolation, polypropylene tubes and endotoxin-free plastic ware are recommended as glass and polyethylene can cause PMN activation.
(3) All centrifuging steps are set at RT to prevent clumping of PMNs at low temperature (below 10°C).
(4) Pipetting must be done very gently, cushioned along the side of the tube, as harsh pipetting may activate PMN.

3. Functional Assays of PMN Function

3.1 *Detection of surface activation markers and intracellular cytokines by flow cytometry*

3.1.1 *Materials*

(1) BD (Becton Dickinson) Falcon™ Clear 96-well Microtest™ plate U-bottom (Cat # 353077)
(2) BD FACS Lysing Solution (Cat # 349202).
(3) VWR deep 96-well Plate (Cat # 82006-448).
(4) Staining buffer: sterile PBS, 2% FBS, 0.1% sodium azide (Sigma Cat # S-2002). Filter the solution and keep at 4°C.
(5) PFA: Paraformaldehyde (16%). (Electron Microscopy Sciences, Cat # 15710). Dilute with PBS to 4%.
(6) BD Perm/Wash Buffer (Cat # 51-2091KZ).
(7) Ligands: LPS (0.5 μg/mL) or TNF-α (20 ng/mL).
(8) Fluorescently-conjugated monoclonal antibodies directed to desired target.
(9) Brefeldin A (BFA, Sigma # B7651) 1 mg/mL in 100% ethanol, and store at –20°C. Final concentration of 10 μg/mL with cells.

3.1.2 *Methods*

PMN can be stimulated in whole blood or after purification. Sample antibody panels are shown in Table 1.

A. Whole blood or purified PMN stimulation
   (1) Add 200 μl whole blood into a U-bottom 96-well plate.
   (2) Stimulate PMN with ligands such as TNF-α (20 ng/ml) for 15 or 30 min at 37°C. PMN in medium without ligand serves as mock untreated control.
B. Surface Labeling. If labeling in whole blood, follow steps 3–7; if using purified PMN, go to step 8. Fluorescently-conjugated antibody labeling should be protected from light.
   (1) Transfer cells to a deep 96-well plate.

Table 1.   Representative Antibody Panels for Detection of PMN Stimulation and Apoptosis by Flow Cytometry. Representative antibody panel choices to detect activation and apoptosis, respectively, with sample wavelengths and fluorochromes indicated. Each panel is also labeled in parallel samples with isotype controls corresponding to the primary antibodies (e.g, Ms IgG1, κ1 Ms IgG2a, κ). PI, propidium iodide.

		Blue Laser 488 nm			Red Laser 633–635 nm		Violet Laser 405 nm
	Filters	530/30	575/26	682/33	660/20	780/60	440/40
Panel	Fluoro-chrome	FITC	PE	PerCP	APC	APC-Cy7	Pacific Blue/eFluor 450
Activation		CD18	IL-8	CD15	CD11b	CD62L	TNFα
Apoptosis		Annexin V	PI	CD15			

(2) Add 2.0 mL 1 × lysing solution to each well to lyse erythrocytes. Vortex gently.

(3) Incubate the plate (protect from light) at RT for 15 m.

(4) Centrifuge the plate at 250 × g for 10 m. Aspirate supernatant.

(5) Wash the cells by adding 1mL 1 × PBS, centrifuge the plate at 250 × g for 10 m.

(6) Aspirate supernatant.

(7) Antibody labeling (protect from light). Re-suspend cells in 50 μL staining buffer (4°C) containing antibody cocktail for cell lineage markers and surface activation markers. In parallel wells, isotype controls for primary antibodies as required (Table 1). Incubate the plate at 4°C in the dark for 30 m.

(8) Add cold (4°C) staining buffer (150 μL/well) to stop binding. Centrifuge at 250 × g for 10 m at 4°C. Aspirate supernatant.

(9) Re-suspend cells in 100 μL 4% PFA for fixing. Incubate at RT for 15 m protected from light. If only surface labeling is planned, stop here. Cells are held in staining buffer until analysis by flow. Centrifuge at 500 × g for 10 m at 4°C.

C. Intracellular staining for cytokine detection (protect from light).

(1) Permeabilization of cells. Aspirate supernatant. Add 1×
    Perm/Wash Buffer to the pellet (200 μL/well), resuspend,
    and centrifuge at 500 × g for 10 m at 4°C. Aspirate superna-
    tant. Re-suspend cells in 200 μL 1 × Perm/Wash Buffer on ice
    and incubate for 15 m. Centrifuge at 500 × g for 10 m at 4°C.
(2) Antibody labeling. Remove the supernatant. Thoroughly
    resuspend fixed/permeabilized cells in 1× Perm/Wash
    Buffer (50 μL/well) containing detection antibodies (Table
    1). Incubate at 4°C for 30 m protected from light.
(3) Wash wells with 1 × Perm/Wash Buffer (150 μL/well), cen-
    trifuge at 500 × g for 10 m at 4°C. Remove the supernatant.
(4) Re-suspend in 200 μL staining buffer (4°C) for sample acqui-
    sition by flow cytometry.

3.1.3 *Notes*

(1) Immediate analysis of cells on flow cytometer is preferred.
    However, fixed cells may be analyzed the following day.
(2) Single-stained cells or compensation beads (eBioscience) are
    used for compensation controls prior to running samples.
(3) The concentration of the isotype control antibody should be the
    same as the specific monoclonal antibody.
(4) Data is analyzed using Flowjo or BD FACS Diva software.

3.2 *Analysis of PMN apoptosis by flow cytometry*

3.2.1 *Materials*

(1) 96-well U-bottom plate.
(2) Stimulants: TNF-α, LPS
(3) PBS (4°C)
(4) FITC- Annexin V Apoptosis Detection Kit I (BD Pharmingen Cat
    # 556547)
(5) Components: 10 × Annexin V Binding Buffer- AVBB (0.1 M
    HEPES/NaOH (pH 7.4), 1.4 M NaCl, 25 mM $CaCl_2$). Dilute to
    1 × with distilled water. FITC- Annexin V (5 μL/test) Propidium
    iodide (PI) — 5 μg/mL prepared fresh daily.

3.2.2 *Methods*

(1) Incubate 1× 10^5 purified PMN for 18 hrs in 100 μL RPMI with 10% human serum with ligands, e.g. TNFα (20 ng/mL), LPS (0.5 μg/mL). Cells in medium without ligands serve as a control.

(2) Stop stimulation by adding PBS (4°C) and washing the cells twice at 250 × g for 10 m.

(3) Aspirate supernatant and resuspend the pellet in 1× binding buffer (10^5 cells/100 μL).

(4) To 1× 10^5 cells/100 μL in a 5 mL tube add 5 μL FITC- Annexin V and 4 μL PI.

(5) Incubate for 15 m at RT protected from light.

(6) Stop incubation by adding 1× binding buffer (400 μL) to each tube.

(7) Immediately assess labeling by flow cytometry.

(8) Express data as a two-color dot plot with FITC-Annexin V and PI.

3.2.3 *Notes*

• Sample antibody panel is shown in Table 1.
• Apoptosis is a continuing process, thus cells should be assessed promptly after staining with Annexin V.

3.3 Detection of specific protein in PMNs by immunoblot

3.3.1 *Materials*

(1) Standard SDS-PAGE gel apparatus and buffers (Invitrogen)

(2) Standard transfer apparatus and buffers (Invitrogen)

(3) Lysis buffer: CelLytic M Cell Lysis Reagent (Sigma, C2978), Protease Inhibitors Tablet (Roche, 11 837 580 001), 1mM NaF, 1mM Na_3VO_4, and 1mM PMSF
Sample buffer: 4× NuPAGE LDS sample buffer (Invitrogen, NP0007) with Sample Reducing Agent (Invitrogen, NP0009).

(4) BCA Protein Assay Kit (Thermo Scientific)

(5) Wash Buffer: Tris-Buffered Saline with 0.1% Tween20 (TBST)

(6) Blocking Buffer: 5% BSA or No-Fat milk in TBST

(7) PVDF membrane (Bio-Rad, 162-0177)

(8) Enhanced Chemiluminescence (ECL) Substrate (Bio-Rad, 170-5060)

3.3.2 *Methods*

(1) Isolate PMN and stimulate according to study protocols. Wash in PBS by centrifugation at $250 \times$ g, 10 m, RT.

(2) Prepare protein samples from freshly isolated or treated PMNs (1–10×10^6 cells$/125$ μL) by lysing washed cells with lysis buffer on ice for 30 m followed by centrifugation ($18{,}000 \times$ g, 30 m, 4°C). Collect supernatant. Determine protein concentration by BCA protein assay.

(3) Mix cell lysate with $4 \times$ sample buffer and heat at 70°C for 10 m. Load 15–20 μg of protein per lane and separate proteins on a NuPAGE Novex Bis-Tris 4–12% Gel at 100 mA for 2–3 hrs.

(4) Transfer proteins from gel to PVDF membrane at 40 V for 2–3 hrs on ice or 14 V overnight. Block non-specific antibody sites on PVDF membrane with blocking buffer on a rocking platform for 1 hr at RT followed by washes (5 m $\times$ 3).

(5) Probe proteins with primary antibody in blocking buffer with 0.05% sodium azide on a rocking platform at 4°C overnight followed by washes (5 m $\times$ 3). Detect bound antibody by HRP conjugate secondary Ab on a rocking platform for 45 m at RT followed by washes (5 m $\times$ 3). Visualize proteins by exposing the membrane to the mixed ECL solutions for 1 m at RT.

3.3.3 *Notes*

• Keep PMN lysate on ice at all times during sample preparation to avoid protein degradation. Aliquot cell lysate and store at −20°C. Avoid repeated thawing.

3.4 *Detection of gene expression in PMN by quantitative PCR (Q-PCR)*

3.4.1 *Materials*

(1) Qiagen RNeasy Mini Kit (74104).
(2) cDNA templates with known concentration of gene of interest.
(3) A pair of primers specific for the genes of interest and internal control (housekeeping gene, e.g. β-actin, GAPDH.
(4) iScript cDNA Synthesis Kit (Bio-Rad, 170-8891).
(5) SsoFast EvaGreen Supermix (Bio-Rad, 172-5200).
(6) 96-well PCR plates (Bio-Rad, HSS-9665).

3.4.2 *Methods*

(1) Extract RNA from freshly isolated or cultured PMNs using Qiagen RNeasy Mini Kit. Determine RNA concentration and quality by Nanodrop. High quality RNA will have an $OD_{260/280}$ ratio of 1.8–2 and an $OD_{260/230} \geq 1.8$. For extraction of micro-RNA, use Qiagen miRNeasy Mini Kit (217004).
Synthesize first-strand DNA using iScript cDNA Synthesis Kit.
(2) Prepare standards and cDNA samples for qPCR using 10-fold serial dilutions of a cDNA template with a known concentration of gene of interest (e.g. β-actin) as standards. Dilute cDNA samples to 1:10 dilution with Nuclease-free water.
(3) Prepare master mix using 2× EvaGreen Supermix and primers (0.2 μM final concentration each). Add 15 μL of master mix and 5 μL each of standard, samples, and negative control (no cDNA) to the 96-well plate. Mix gently by pipetting up and down. Perform qPCR using Real-Time PCR machine (C1000-Thermal Cycler, Bio-Rad).
(4) Analyze data by checking the standard curve for a linear standard curve ($R^2 > 0.980$) and high amplification efficiency (90–102%). Absolute quantification is obtained by calculating the copy number of the unknown samples based on the equation of the linear regression line [Ct = m (log quantity) + b]: Quantity = $10^{(Ct-b)/m}$. Relative quantification is obtained by $2^{-\Delta\Delta Ct}$ method (14).

3.4.3 Notes

• It is recommended to perform first-strand DNA synthesis immediately after RNA isolation to avoid RNA degradation. However, RNA samples can be kept overnight at 4°C and stored for a longer time at –80°C.

3.5 Detection of PMN secretion of cytokines or chemokines by ELISA

3.5.1 Materials

(1) 96-well ELISA plates (BD Falcon, 353279)
(2) Antigen-specific capture antibody and HRP-conjugated detection antibody for target of interest or corresponding pairs for multiplex kit.
(3) Recombinant cytokine target of interest for titration of standard curve; aliquot after reconstitution and store at –80°C
(4) ELISA kit buffers and detection enzyme substrates: Coating buffer: 0.1 M sodium carbonate, pH 9.5; Assay Diluent: PBS with 10% FBS, pH 7.0; Wash buffer: PBS with 0.05% Tween-20; Substrate A: hydrogen peroxide; Substrate B: tetramethylbenzidine, TMB; Stop solution: 2N sulfuric acid.

3.5.2 Methods

(1) Collect supernatant from PMN after stimulation protocol, centrifuge at $250 \times g$, 10 m, RT. Assay immediately or store samples at –20°C until batch assay.
(2) ELISA kit protocol: Coat plates with optimized concentrations of capture antibody (100 µL/well) at 4°C overnight. Block with assay diluent (200 µL/well) at RT for 1 hr. Add 2-fold serial dilutions of standard of interest in assay diluent, sample, and control (100 µL/well) and incubate at RT for 2 hr. Add 100 µL of detection antibody (100 µL/well) and incubate at RT for 1 hr. Pre-mix substrate A and B in equal volumes within 15 m prior to use and protect from light. Add 100 µL of substrate solution into each well and incubate at RT for 30 m in the dark. Add stop solution (50 µL/well) and read absorbance at 450 nm within 30 m. For

wavelength correction, set to 540 nm or 570 nm. The corrected readings = $OD_{450} - OD_{540}/_{570}$.

3.5.3 *Notes*

(1) Bring all reagents and samples to room temperature before use.
(2) Samples may require dilution to fall into the range of working standards.
(3) Complete removal of liquid at each step is essential to good performance; remove remaining buffers by inverting the plate and tapping it against clean paper towels.

4. Multidimensional Profiling of PMN Immune Status

The assays presented here provide quantitative measures of different components of PMN functions. For a more in-depth assessment of PMN function, several of these assays can be conducted in parallel, provided the number of samples per day is not excessive. An outline of such analysis (Fig. 1) shows multiple wells of PMN beginning stimulation protocols together for harvest at short time intervals (15, 30 m) for flow cytometry of surface activation markers (Figure 1b) and for protein lysate for immunoblot of signaling intermediates (Fig. 1c). Parallel samples under the same treatment conditions would be incubated for longer times (4 hrs) for RNA collection for qPCR (Fig. 1d) or to allow secreted cytokines to accumulate for ELISA assays (Fig. 1e) or even overnight for assessment of apoptosis (Fig. 1f). This multidimensional profiling of PMN from the same subject is generally feasible for human samples as the cell number is abundant. This requires care in planning and speed in execution but generates a rich data set for analysis. Indeed, multiple components of analysis from the same sample supports more in depth understanding of the cell status and contribution to the condition under study.

Acknowledgements

This work was supported in part by the National Institutes of Health (N01   HHSN272201100019C, U19AI089992).   The   authors   are grateful to many colleagues and former lab members for development of these protocols.

References

1. Jaillon S, Galdiero MR , Del Prete D *et al.* (2013) Neutrophils in innate and adaptive immunity. *Semin Immunopathol* **35**: 377–394.
2. Nauseef WM, Borregaard N. (2014). Neutrophils at work. *Nat Immunol* **15**: 602–611.
3. von Vietinghoff S, Ley K. (2008) Homeostatic regulation of blood neutrophil counts. *J Immunol* **181**: 5183–5188.
4. Amulic B, Cazalet C, Hayes GL *et al.* (2012) Neutrophil function: From mechanisms to disease. *Ann Rev Immunol* **30**: 459–489.
5. Papayannopoulos V, Zychlinsky A. (2009) NETs: A new strategy for using old weapons. *Trends Immunol* **30**: 513–521.
6. Mocsai A. (2013) Diverse novel functions of neutrophils in immunity, inflammation, and beyond. *J Exp Med* **210**: 1283–1299.
7. Bowers NL, Helton ES, Huijbregts RP *et al.* (2014) Immune suppression by neutrophils in HIV-1 infection: Role of PD-L1/PD-1 pathway. *PLoS Pathog* **10**: e1003993.
8. Guo X, Booth CJ , Paley MA *et al.* (2009) Inhibition of neutrophil function by two tick salivary proteins. *Infect Immun* **77**: 2320–2329.
9. van Eeden SF, Klut ME, Walker BA *et al.* (1999) The use of flow cytometry to measure neutrophil function. *J Immunol Methods* **232**: 23–43.
10. Brandt E, Woerly G, Younes AB *et al.* (2000) IL-4 production by human polymorphonuclear neutrophils. *J Leukoc Biol* **68**: 125–130.
11. Wittmann S, Roth G, Schmitz *et al.*(2004) Cytokine upregulation of surface antigens correlates to the priming of the neutrophil oxidative burst response. *Cytometry A* **57**: 53–62.
12. Qian F, Guo X, Wang X *et al.* (2014) Reduced Bioenergetics and Toll-like Receptor 1 function in human polymorphonuclear leukocytes in Aging. *Aging* **6**: 131–139.
13. Bai F, Kong KF, Dai J *et al.* (2010) A paradoxical role for neutrophils in the pathogenesis of West Nile virus *J Infect Dis* **202**: 1804–1812.
14. Livak KJ, Schmittgen TD. (2001) Analysis of relative gene expression data using real-time quantitative PCR and the 2(-Delta Delta C(T)) Method. *Methods* **25**: 402–408.

Chapter 9

Multiplexed Transcriptomic Profiling Using Color-Coded Probe Pairs

Adrian K. Wyllie* and Jose D. Herazo-Maya*,†

*Section of Pulmonary, Critical Care and Sleep Medicine
Yale University, School of Medicine, New Haven
CT, P.O. 06510 USA
†jose.herazo-maya@yale.edu

Abstract

The completion of the Human Genome Project created a need for robust high-throughput transcriptomic profiling platforms. A comprehensive knowledge of transcript abundance will help us to understand the molecular infrastructure of a disease process which can ultimately guide therapy. Real Time Polymerase Chain Reaction (RT-PCR), Microarrays and RNA sequencing are among the different gene expression profiling technologies available for transcriptomic analysis. In this chapter, we describe a novel, state-of-the-art profiling technology based on the principle of color-coded probe pairs, which measures RNA abundance in a multiplexed fashion with the capability to overcome the limitations of other current profiling platforms.

---

†Corresponding author.

1. Introduction

The central dogma of molecular biology states that the flow of genetic information originates with the transcription of DNA to messenger RNA (mRNA) which is subsequently translated to a protein. Albeit correct, we now have learned that there are exceptions to this rule given the presence of several regulatory transcripts which can alter the expression of multiple genes. A major milestone in our understanding of gene expression can be attributed to the Human Genome Project. Decoding the human genome revolutionized the field of human genetics and biomedical research, helping us to discover new regulatory transcripts and their functions, in addition to providing us with an extensive knowledge about disease pathogenesis and our inherited susceptibility. With the completion of the Human Genome Project in 2001, the new field of transcriptomics emerged. Transcriptomics is the field that studies the simultaneous expression of several RNA molecules such as mRNA, microRNAs (miRNAs) and long non-coding RNA (lncRNAs), in a single cell, group of cells and tissue, and their role in disease pathogenesis.

In the last decade, the large majority of transcriptomic studies largely relied on hybridization-based microarray technologies. More recently, the introduction of high-throughput next-generation sequencing (NGS) revolutionized the field of transcriptomics by allowing RNA analysis through sequencing at massive scale (RNA-Seq).[1] Here, we introduced the principle of transcript count using color-coded probe pairs (nCounter® Analysis System, Nanostring) which has been employed in life sciences research since its introduction in 2008. This high-throughput expression profiling technology is a direct and reproducible digital detection system which enables targeted analysis of multiple RNA transcripts.

2. The Color-Coded Probe Pair Principle and Applications

The nCounter analysis system platform is based on direct measurement of transcript abundance by using multiplexed, color-coded probe pairs.[2] This system is based on the hybridization of two probes (the Capture Probe and the Reporter Probe) with the transcript of interest. The Capture Probe (Fig. 1a), contains a 35- to 50-base sequence complementary to the transcript of interest (Fig. 1b) plus a short sequence coupled with a biotin tag at the 3' end. The Reporter Probe (Fig. 1c), contains a second 35- to 50-base sequence, also complementary to the targeted transcript, coupled to a color-coded tag at the 5' end. A hybridization step, allows the formation of target-probe complexes (Fig. 1d). The color-coded tags are single-stranded DNA molecules coupled to a series of complementary transcribed RNA segments, each one labeled with a specific fluorophore. The order of the color-coded tags in the target-probe complexes creates a unique code for each transcript of interest, providing a detection signal that is captured and counted by a high sensitivity scanner (Fig. 1e). An additional step, the ligation step, is required to capture miRNA targets and increase the sensitivity of the standard nCounter assay. In this step, *in vitro* synthetized DNA sequences (miRtags) are ligated to complementary, mature miRNAs, with sequence-specific oligonucleotides bridges (Fig. 2a). After the ligation step is performed, the remaining steps such as hybridization (Fig. 2b) and transcript count (Fig. 2c) are similar to the standard nCounter assay.

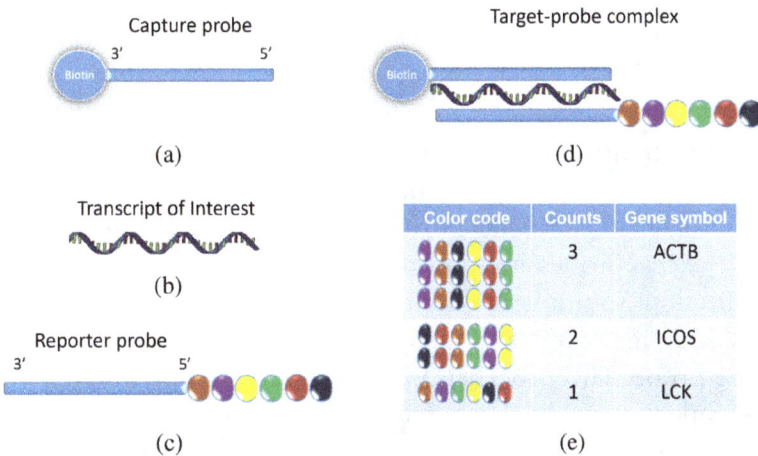

Fig. 1. Overview of the Color-Coded Probe Pair Principle (a) Capture probe with an attached Biotin label at the 3' end (b) Transcript of interest (b) Reporter probe carries a unique color code at the 5' end (d) Target-probe complex formed by the capture probe, the transcript of interest and the reporter probe after hybridization (e) Transcript count is performed based on the number of color codes per transcript.

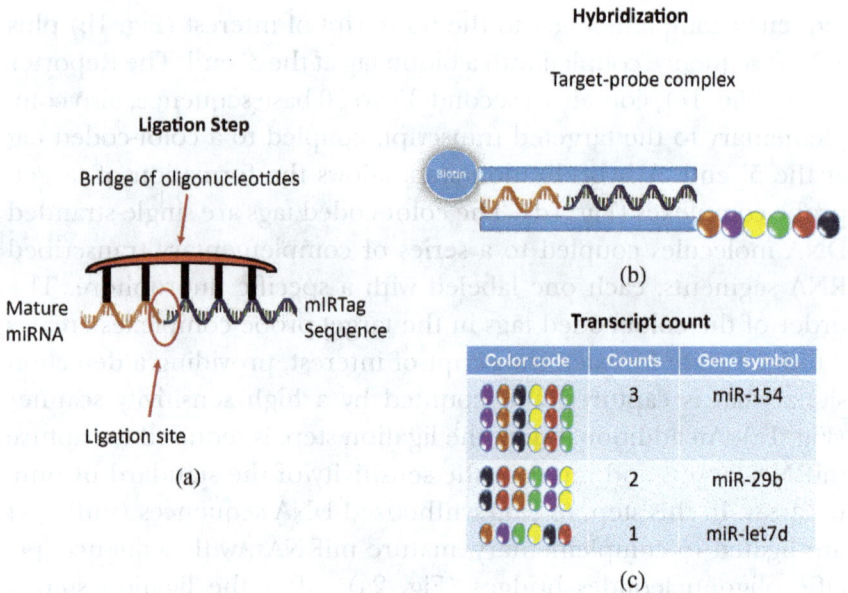

Fig. 2. miRNA Sample Preparation for the nCounter Analysis System. (a) DNA sequences called miRtags are ligated to mature miRNAs through complementary, sequence-specific bridging oligonucleotides. (b) The miRtagged mature miRNA is then hybridized to a probe pair in the standard nCounter gene expression assay workflow. (c) Transcript count is performed based on the number of color codes per miRNA transcript.

The nCounter analysis system consists of two instruments, the Prep Station and the Digital Analyzer. The Prep Station is used to perform a magnetic bead purification step after hybridization, to remove excess probes and non-target cellular transcripts. Briefly, the hybridization solution containing the target-probe complexes is mixed with magnetic beads complementary to sequences on the capture probe. This process is followed by a wash step to remove the excess reporter probes. The target-probe complexes are then hybridized to magnetic beads complementary to sequences on the reporter probe. A final wash step is performed to remove excess capture probes. Lastly, the purified target-probe complexes are eluted off the beads and immobilized on the cartridge for transcript count measurement and data collection in the Digital Analyzer.

The nCounter analysis system can measure up to 800 different RNA targets simultaneously from samples such as crude cell and tissue lysates,[3–5] blood, [6]formalin-fixed paraffin embedded (FFPE) tissue [7,8] and single cells.[9,10] Up to 12 samples can be performed per run with a maximum of 36 samples per day. The hybridization step is usually performed in less than an hour and the remaining steps are performed by the Prep Station and the Digital Analyzer. The full process can be completed in 24 hours.

There are several factors to consider when selecting the nCounter analysis system over other commonly used transcriptomic analysis platforms. The nCounter system is typically used by researchers as a way to perform validation of data obtained by genome-wide transcriptomic profiling techniques such as Microarrays and RNA-sequencing. Among the benefits of using the nCounter system for high-throughput validation over Microarrays and RNA-Seq, is that it avoids the complex experimental process and data analysis necessitated by these two platforms. Furthermore, the system bypasses the need for amplification and reverse transcription required for RT-PCR, even for low abundance transcripts. Additionally, the nCounter has generated superior gene expression quantification results with formalin-fixed paraffin embedded tissue (FFPE) samples[7] when compared to other methods such as Microarrays or RT-PCR. Table 1 presents a comparison between the nCounter analysis system and other commonly used platforms for transcriptomic profiling.

3. Materials and Methods

3.1 *12 nCounter gene expression hybridizations*

3.1.1 *Materials*

(1) nCounter gene expression (GX) assay kit (provided by Nanostring): Reporter CodeSet, Capture ProbeSet and Hybridization buffer.
(2) miRNeasy Mini Kit (Qiagen).
(3) Disposable gloves.
(4) Nuclease-free water.

Table 1.   Comparison of Available Platforms for Transcriptomic Profiling.

Technical aspects	nCounter analysis system	RNA-Seq	Microarray	qRT-PCR
Input total RNA	10–100 ng	>100 ng	50–200 ng	>5 ng
Throughput time	~24 hrs	~3 days	~2–3 days	1.5–3 hrs
Approximate cost per sample	$80–160	400–800+	200–400+	20–50+
Advantages	• Low starting RNA input. • No reverse transcriptase or amplification step required. • Single reaction per sample. • Good performance with degraded RNA.	• Can detect novel transcripts. • Can identify Single Nucleotide Polymorphisms (SNPs). • Free of background hybridization. • Good performance with degraded RNA.	• Low starting RNA input. • Mature informatics and statistics. • High reproducibility.	• High specificity and sensitivity. • Large dynamic range. • No post-amplification processing.
Disadvantages	• Special equipment is required. • Selection of multiple endogenous controls is required for data normalization.	• High starting RNA input. • Labor intensive. • Complex data processing. • Expensive.	• Limited dynamic range. • Labor intensive. • Poor performance with degraded RNA.	• Need for reverse transcriptase step. • High RNA input required for high-throughput experiments. • False positives due to high sensitivity.

3.2 *Instruments*

(1) NanoDrop ND-1000.
(2) Analyzer of RNA integrity (2200 TapeStation or similar).
(3) Pipette for 0.5–10 μL.
(4) Pipette for 2–20 μL.
(5) Pipette for 20–200 μL.
(6) Tube-Strip Picofuge.
(7) DNA engine thermocycler, hybridization oven (or similar).

3.3 *Methods*

3.3.1 *Hybridization Step*

(1) Extract total RNA using the miRNeasy Mini Kit (Qiagen) following the manufacturers protocol.
(2) Dilute 100 ng of total RNA in 5 μl of nuclease-free water for a concentration of 20 ng/μL.
(3) Remove aliquots of both Reporter CodeSet (Green cap) and Capture ProbeSet (Silver cap) reagent from the freezer and thaw on ice. Invert at least ten times to mix well and spin down. Do not vortex.
(4) Add 130 μL of hybridization buffer to the Reporter CodeSet tube to create a master mix. Do not remove the Reporter Probes from the tube. Invert to mix and spin down. Do not mix by pipetting.
(5) Label a 12 tube strip (provided by Nanostring) and cut it in half so it will fit in a picofuge.
(6) Add 20 μL of the master mix to each of the 12 tubes. It is recommended to use a fresh tip for each pipetting step.
(7) Add 5 μL of the diluted RNA sample to each tube (100 ng of total RNA). The final volume for this step should not be more than 25 μL.
(8) Pre-heat the thermocycler to 65°C using a volume of 30 μL, a heated lid and "forever" time setting. A 65°C hybridization oven may be used if a thermocycler is not available.
(9) Add 5 μL of the capture ProbeSet solution to each tube immediately before placing at 65°C. Cap tubes and mix the reagents by inverting the strip tubes at least 10 times. Do not mix by

pipetting. Briefly spin down and immediately place the strip tube in the 65°C thermocycler.

(10) Incubate hybridization assays for at least 12 hrs. Hybridizations should be left at 65°C until ready for processing. Once removed from the thermocycler, proceed immediately to the magnetic bead purification step in the nCounter Prep Station.

4. 12 nCounter miRNA Expression Hybridizations

4.1 *Materials*

(1) nCounter miRNA expression assay kit (provided by Nanostring): miRNA Assay Controls, Annealing buffer, miRNA Tag Reagent, PEG, Ligase Buffer, Ligase, Ligation Clean-Up Enzyme, Reporter Code Set, Capture Probe Set and Hybridization buffer.
(2) miRNeasy Mini Kit (Qiagen).
(3) Disposable gloves.
(4) Nuclease-free water.

4.2 *Instruments*

(1) NanoDrop ND-1000.
(2) Analyzer of RNA integrity (2200 TapeStation or similar).
(3) Pipette for 0.5–10 μL.
(4) Pipette for 2.0–20 μL.
(5) Pipette for 20–200 μL.
(6) Picofuge with strip-tube adaptor.
(7) DNA engine thermocycler, hybridization oven (or similar).

4.3 *Methods*

4.3.1 *Ligation Step*

(1) Extract total RNA using the miRNeasy Mini Kit (Qiagen), following the manufacturers protocol.
(2) Dilute 100 ng of total RNA in 35 μL of nuclease-free water for a concentration of 33 ng/μL.

(3) Prepare a 1:1500 dilution of the miRNA assay controls. This can be performed by adding 499 μL of nuclease-free water to 1 μL of the miRNA assay control in a sterile microfuge tube. Vortex and store on ice.

(4) Prepare an annealing master mix by combining 13 μL of Annealing Buffer, 26 μL of nCounter miRNA Tag Reagent and 6.5 μL of the 1:1500 miRNA Assay Control dilution prepared in Step 3. Mix well by pipetting up and down.

(5) Aliquot 3.5 μL of the annealing master mix into each tube of a 12 × 0.2 mL strip tube.

(6) Add 100 ng of RNA sample diluted in 3 uL to each tube. Cap tubes and flick tubes gently to mix. Place strip in thermocycler and initiate the Annealing Protocol (94°C for 1 m, 65°C for 2 m, 45°C for 10 m, 48°C hold).

(7) Combine 19.5 μL of PEG and 13 μL Ligation Buffer to prepare a ligation master mix.

(8) Following completion of the Annealing Protocol, when the thermocycler has reached 48°C, add 2.5 μL of the ligation master mix to each tube.

(9) Return tubes to 48°C thermocycler, close lid, and incubate at 48°C for 5 m.

(10) Open thermocycler, carefully remove caps from tubes, leaving strip in place in the heat block, and add 1.0 μL of Ligase directly to each tube while incubating at 48°C.

(11) Immediately after addition of Ligase to the final tube, recap tubes (leaving tubes in heat block), close thermocycler, and initiate Ligation Protocol (48°C for 3 minutes, 47°C for 3 m, 46°C for 3 m, 45°C for 5 m, 65°C for 10 m, 4°C hold).

(12) After completion of Ligation Protocol, add 1 μL Ligation Clean-Up Enzyme to each reaction. The tubes can be removed from the heat block for this step. Initiate Purification Protocol (37°C for 1 hr, 70°C for 10 m, 4°C hold).

(13) After completion of Purification Protocol, add 40 μL of RNAse-free water to each sample. Mix well and spin down. Be sure to denature (see miRNA hybridization) before adding prepared sample in the hybridization protocol.

4.3.2 *Hybridization Step*

(1) Remove aliquots of both Reporter CodeSet (Green cap) and Capture ProbeSet (Silver cap) reagent from the freezer and thaw on ice. Invert at least ten times to mix well and spin down. Do not vortex.

(2) Add 130 μL of hybridization buffer to the Reporter CodeSet tube to create a master mix. Do not remove the Reporter Probes from the tube. Mix well.

(3) Label a provided 12 tube strip and cut it in half so it will fit in a picofuge. Add 20 μL of master mix to each of the 12 tubes.

(4) Denature samples from the miRNA sample prep protocol at 85°C for 5 m and quick-cool on ice. Add a 5 μL aliquot from the miRNA Sample Preparation Protocol to each tube.

(5) Pre-heat the thermocycler to 65°C using a volume of 30 μL, a heated lid and "forever" time setting. A 65°C hybridization oven may be used if a thermocycler is not available.

(6) Add 5 μL of Capture ProbeSet to each tube immediately before placing at 65°C. Cap tubes and mix the reagents by inverting the strip tubes several times and flicking with your finger to ensure complete mixing. Briefly spin down at <1000 rpm and immediately place the strip tube in the 65°C thermocycler.

(7) Incubate hybridization assays for at least 12 hrs. Hybridizations should be left at 65°C until ready for processing. Once removed from the thermocycler, proceed immediately to the magnetic bead purification step with the nCounter Prep Station.

5. Magnetic Bead Purification

5.1 *Materials*

(1) Hybridization assays

5.2 *Instruments*

(1) nCounter Prep-station.
(2) nCounter Digital Analyzer.
(3) Memory stick.

(4) The following instruments are provided by Nanostring for each set of 12 reactions:

(5) Reagent plates (#2).

(6) Cartridge (#1).

(7) Tip box (#1).

(8) Tip sheaths (#2).

(9) Tube strips (#2).

(10) Self-adhesive paper (#1).

5.3 Methods

(1) Remove the sample cartridge from the freezer and reagent plates from the fridge.

(2) Allow the cartridge and plates to reach room temperature before using.

(3) Centrifuge the reagent plates at 2000g for 2 m.

(4) Place the following instruments on the Prep Station deck:

- Sample tubes.
- Reagent plates.
- Tips.
- Tip sheaths.
- Empty strip tubes.

(5) Close the tube holder and secure the electrode fixture over the sample cartridge on the Prep Station deck.

(6) Start that Prep Station.

(7) Remove the cartridge from the Prep Station deck once the purification step is completed. Cover the cartridge wells with self-adhesive paper and transfer it to the Digital Analyzer.

(8) Start Digital Analyzer.

(9) Transfer the data files to a memory stick.

(10) Proceed to data normalization and analysis.

6. Data Normalization

Data normalization can be performed using the nSolver, a user friendly analysis software provided by Nanostring. The data files produced by the nCounter Digital Analyzer are exported as a Reporter Code Count (RCC) file. RCC files are comma-separated text (.csv) files that contain the counts for each transcript of interest in a sample. These files are used for data normalization.

Two types of normalizations are required prior to data analysis: positive control normalization and housekeeping gene normalization. Other normalization methods can be performed and they are available in the NanoStringNorm R package.[11] We will discuss here the positive control and housekeeping gene normalizations, which can be performed using the nSolver Analysis Software.

Positive control normalization can be used to normalize all sources of variation associated with the platform. Typically, these variations are introduced during hybridization and magnetic bead purification steps. This normalization is performed by using positive controls which are synthetic RNA targets that have been spiked in to the CodeSet at concentrations ranging from 0.1fM–100fM. Each reaction contains a specific set of positive controls. The average of the positive control values across all experiments is used as the reference against which each individual experiment is normalized.

Reference gene normalization can be performed to adjust counts of all transcripts relative to a set of transcripts that are not expected to significantly change among samples or replicates. The selection of the housekeeping genes is essential for this type of normalization because differences in expression of housekeeping genes among tested groups can significantly alter the results of the experiment. This is why it is recommended to add at least three endogenous controls (and potentially more) to the custom-made CodeSets. The housekeeping gene normalization is performed by using the geometric mean of the reference genes for each experiment to calculate scaling factors. The average of these geometric means across all experiment is then used as the reference against which each experiment is normalized.

Once these two types of normalizations have been applied to the raw data, different statistical tests can be applied to identify significant differences among tested groups.

References

1. Ozsolak F, Milos PM. (2011) Rna sequencing: Advances, challenges and opportunities. *Nat Revs Genet* **12**: 87–98.
2. Geiss GK, Bumgarner RE, Birditt B *et al.* (2008) Direct multiplexed measurement of gene expression with color-coded probe pairs. *Nat Biotechnol* **26**: 317–325.
3. Ernst J, Kheradpour P, Mikkelsen TS *et al.* (2011) Mapping and analysis of chromatin state dynamics in nine human cell types. *Nature* **473**: 43–49.
4. Payton JE, Grieselhuber NR, Chang LW *et al.* (2009) High throughput digital quantification of mrna abundance in primary human acute myeloid leukemia samples. *J Clin Invest* **119**: 1714–1726.
5. Northcott PA, Shih DJ, Remke M *et al.* (2012) Rapid, reliable, and reproducible molecular sub-grouping of clinical medulloblastoma samples. *Acta neuropathologica* **123**: 615–626.
6. Yamamoto M, Singh A, Ruan J *et al.* (2012) Decreased mir-192 expression in peripheral blood of asthmatic individuals undergoing an allergen inhalation challenge. *BMC genomics* **13**: 655.
7. Reis PP, Waldron L, Goswami RS *et al.* (2011) Mrna transcript quantification in archival samples using multiplexed, color-coded probes. *BMC Biotechnol* **11**: 46.
8. Kojima K, April C, Canasto-Chibuque C *et al.* (2014) Transcriptome profiling of archived sectioned formalin-fixed paraffin-embedded (as-ffpe) tissue for disease classification. *PloS one* **9**(1):e86961.
9. McDavid A, Dennis L, Danaher P *et al.* (2014) Modeling bi-modality improves characterization of cell cycle on gene expression in single cells. *PLoS Comp Biol* **10**(7): e1003696.
10. Guo G, Luc S, Marco E *et al.* (2013) Mapping cellular hierarchy by single-cell analysis of the cell surface repertoire. *Cell Stem Cell* **13**: 492–505.
11. Waggott D, Chu K, Yin S *et al.* (2012) Nanostringnorm: An extensible r package for the pre-processing of nanostring mrna and mirna data. *Bioinformatics* **28**: 1546–1548.

Chapter 10

Statistical Analysis of Human Immunologic Studies: Mixed Effects Modeling

Heather G. Allore[*,†,‡,§] *and Mark Trentalange*[†,¶]

[†]*Yale University Program on Aging, Department of Medicine*
300 George Street — Suite 775, New Haven
Connecticut 06511, USA

[‡]*Department of Biostatistics, Yale School of*
Public Health, 60 College Street
P.O. Box 208034, New Haven, CT 06520-8034, USA

[§]*heather.allore@yale.edu*

[¶]*mark.trentalange@yale.edu*

1. Introduction

One of the most striking differences in translational research is the larger degree of variability among human subjects compared to animal models. Because lifetime exposures are accumulated, variability is often further heightened in studies of the elderly. For example, almost three quarters of persons 65 years and older have multiple chronic diseases.[1] Thus, while we would expect to see significant changes in

---

*Corresponding author.

the innate immune system in older adults, we need efficient unbiased methods that can address these additional sources of variability. This chapter provides some practical approaches to the statistical analysis of immunologic data from studies with human subjects. Examples[a] of analyzing repeated observations on subjects are provided; first with cross-sectional and then with longitudinal observations.

2. Repeated Observation Studies

2.1 Data visualization

In the exploratory phase raw data should be visualized whenever possible, not only to assess accuracy of experimental procedures and verify data recording and transcription, but also to determine central tendency and variability.

Typically, distribution assessment is done with box-and-whisker plots or histograms (Figs. 1 and 2 top row). The vast majority of regression models assume that the data are normally distributed. This effect may be determined visually by inspecting in the observed data whether the median and mean overlap or whether the 25th and 75th percentiles of the box are equidistant or whether the "whiskers" representing the 95% confidence intervals are equidistant (Fig. 1). Alternatively, overlaying a theoretical normal distribution on a histogram of the observed data (e.g. Fig. 2, top row) provides visual assessment of normality. Using parametric tests for data that are not normally distributed may result in inaccurate and imprecise (biased) conclusions. Biological factors (e.g., amount of cytokines produced, expression of co-stimulatory molecules) may follow a normal distribution; however, several data transformation methods — typically logarithmic or exponential — are available for non-normally distributed data.[3]

Variation describes the spread or scatter of the data, while covariance measures the strength of relationship between two or more variables. Be aware that many commonly used covariance methods assess linear relationships only. If variables are truly independent (i.e., the correlation is zero), the covariance is zero; however, a

---

[a]All figures were produced by SAS® version 9.4 (© SAS Institute, Cary, NC, USA).

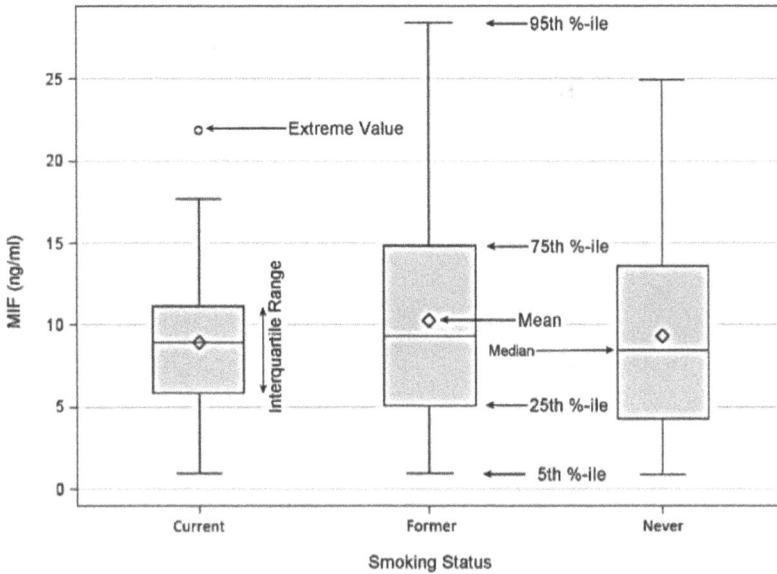

Fig. 1. Box-and-whisker Plot: Macrophage Migration Inhibitory Factor (MIF) levels by smoking status displaying the definitions of box-and-whisker plot elements (data adapted from Sauler *et al.*, (2014)).[2]

strong non-linear relationship (e.g. U-shaped relationship) may also result in a zero covariance. Transformational, non-linear and non-parametric[3] techniques exist for analysis of these situations.

While journals vary in their requirements, we find it helpful to represent measurements as means and 95% confidence intervals. Although model-based adjustments to the data may ensue, this gives the investigator a rough idea of potential differences among groups, experimental conditions, or among time periods.

2.2. Correlations among repeated measures (covariance structures)

While there are several measures of covariance or correlation, the commonly used Pearson's correlation co-efficient assumes two variables are measured on comparable continuous scales and summarizes the extent that two variables are linearly related to each other. Co-efficients range from −1 to 1, with a value near or equal to zero

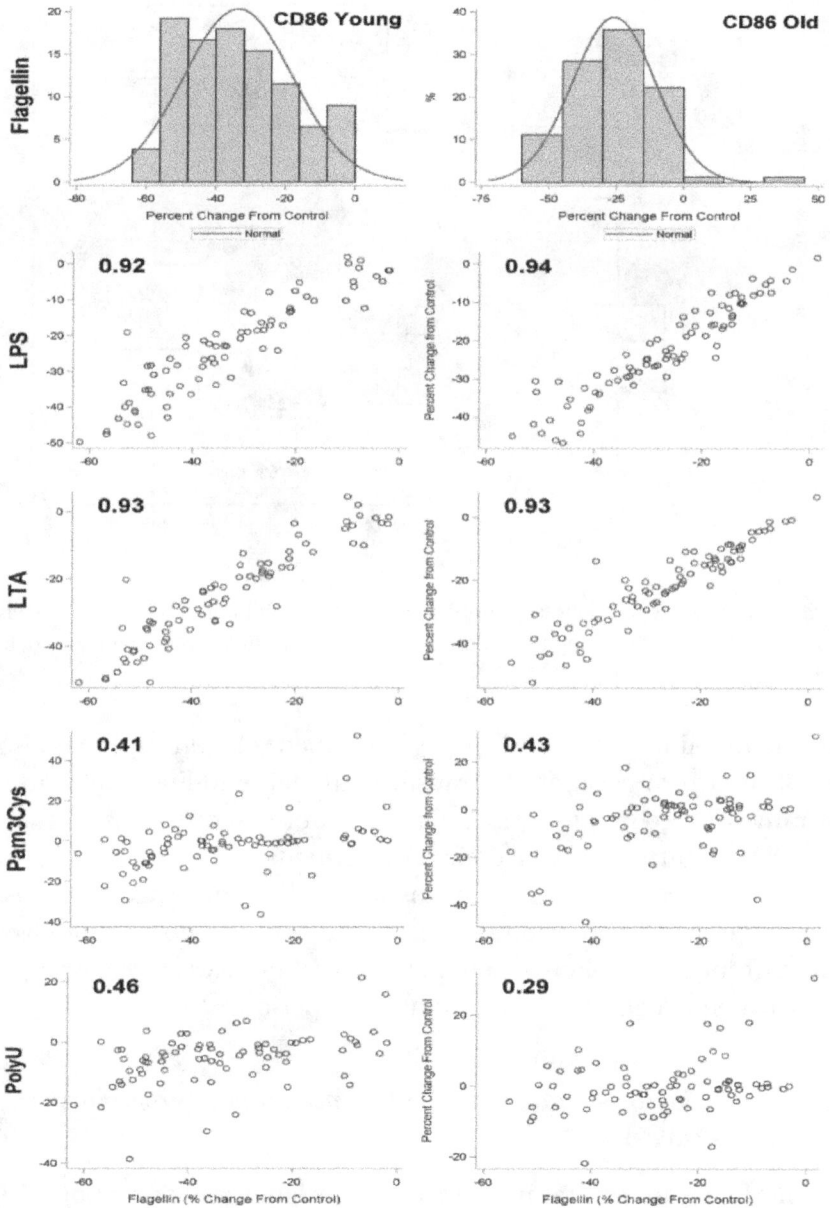

Fig. 2. Histogram and scatter plots: Toll-like Receptor (TLR)–induced CD86 expression in peripheral blood mononucleated cells with various ligands for young and older adults. Histogram plots are for (TLR)–induced CD86 expression with superimposed normal distribution (top row). Scatter plots are for CD86 on cells

Fig. 2.   *(Figure on facing page)*

treated with Flagellin and other ligands (all percent change from unstimulated — control condition) in young (21–30 years old) and older (≥65 years old) adults. The following ligands were added to peripheral blood mononuclear cells: Flagellin, LPS, LTA, Pam3Cys, and PolyU. Numbers in the plot are Pearson's correlation co-efficients (data from van Duin *et al.*, (2005)).[5] Note that the distributions of Flagellin for young adults (top left) poorly fit the normal curve and that the distributions are not similar between young and old adults. In addition, LPS and LTA demonstrate similar strong linear relationships with Flagellin, while Pam3Cys and PolyU have weaker but possibly influential correlations.

implying little or no linear relationship, while values approaching −1 or 1 indicate a strong linear relationship. Creating a correlation matrix among all outcomes (e.g. repeated cytokine expressions) assesses their degree of independence (Fig. 2). Similarly, it is important to measure the correlation among independent variables. Correlation among independent variables violates the model assumption of independence among covariates and results in variance inflation and biased estimates; thus, factors that may be strongly associated with the outcome would not appear to be so.[4]

For any analysis to be valid, the covariance among repeated observations must be modeled properly.[6] It is helpful to consider different covariance structures.[7] Here we will briefly explore four common covariance structures:

Compound Symmetry: This structure assumes that the relationships among repeated outcomes are not independent and that the covariances between all pairs of observations are the same. Repeated measures Analysis of Variance (ANOVA) assumes this covariance structure. This assumption is appropriate when there are only two repeated observations, or the repeated observations arise from the same underlying immunologic mechanism; however, often this assumption does not hold when observations are repeated over time or the underlying mechanisms differ. The data in Fig. 2 shows non-constant correlation among ligands; thus, this would be a poor choice.

Autoregressive: Measurements that are relatively closely spaced (e.g., consecutive measurements at baseline and hours or days later) may be more highly correlated than measurements made farther apart (e.g., baseline and months later). Varieties of autoregressive covariance structures exist (e.g., first-order autoregressive).[7,8]

Independent (or *Variance Components*): Assumes independence, the covariance between repeated measures is zero. This would be a poor choice for data in Fig. 2.

Unstructured: Sometimes no standard covariance specification fits well; an unstructured covariance allows each individual to have its own structure instead of a shared covariance structure. This addresses the problems of a heterogeneous cohort and the cross-sectional repeated measurement of immunologic outcomes. Although it may best model the biology, the disadvantages are that this structure uses the most parameters and may be computationally more demanding (or untenable), as well as more conservative in determining statistical significance. This would be an acceptable choice for the data in Fig. 2.

Not every covariance specification will be appropriate to every analysis. Measurements of different ligands in the same subset of cells (repeated measures — cross-sectional) might best be modeled with an unstructured covariance, whereas repeated measures involving the same ligand over time (longitudinal) might be modeled optimally with an autoregressive structure (measurements closer to each other are probably more similar than measurements distant to one another). There is a balance among parsimony, model fit and ability of the model to converge (come to a stable solution) which determines the final specification of the repeated-measures model. In addition, experimental questions (e.g., characterizing individual versus group differences), type of outcome (categories, counts, percentages, continuous scales) and covariates all play a part in model specifications.

2.3. *Fixed and random variables and effects*

A "fixed variable" is one that has predefined levels or categories, such as gender (male or female) or level (amount of a stimulant) and is repeated among experiments. "Random variables" have assumed representative values drawn from a larger population of values. Most regression models assume the independent variables are fixed. The terms "random" and "fixed" as applied in regression models signify the type of statistical model. Most researchers use fixed effects regression. However, often a sample of human subjects represents only a random sample from a larger population from which they are drawn, such as all persons with rheumatoid arthritis. It is not uncommon for human subjects to have a different level of cytokine expression in unstimulated cells. But after stimulation with a "fixed" variable (e.g., ligand at a certain concentration) they may share the same slope[9] (i.e., change from the unstimulated condition). Employing a random intercept allows each subject to have a potentially unique value for the amount of unstimulated cytokine expression. Analyses using both fixed and random effects are called "mixed effects models."

2.4. Mixed effects models

Mixed effect models are pointedly suited for translational immunologic research, in which the researcher might want to make inferences about fixed effects (e.g., genotype, age groups or disease status) while using heterogeneous human subjects. Including covariates can account for observed sources of variation, but cannot address unmeasured sources of variation.[7, 9] By including random effects the researcher may account for unmeasured factors, such as genetic makeup or environmental exposures, for each subject.

Mixed effects models are useful when analyzing experiments with repeated measurements on the same subject. For translational immunologic research, the application of repeated measurement analysis and random effects make mixed effects models attractive because they can account for the "within subject" correlation of observations which can be unique for each subject. We have employed a mixed effects model to estimate the effects of age on TLR function[5, 10] and have found it improves the overall model fit.

2.5. Generalized estimating equation models

When the outcome is not normally distributed, such as response to strain-specific influenza vaccine, generalized linear models (GEE)[11] or more sophisticated models can be applied. The principles described for mixed models apply, although the distributional form of the outcome (i.e., binary,[12] count,[13] other non-normal distributions[14]) must be specified and underlying estimation procedures differ. Hybrids of these models (generalized linear mixed models, GLMM)), as well as their non-linear analogs exist.

Models should address the hypotheses under investigation; thus, in-depth knowledge of the research questions and the cohort under study is critically important in model creation. The most crucial factor in analyzing data is to select a reasonably parsimonious model appropriate to address the hypothesis.

2.6. Checking model fit

2.6.1. Fit criteria

For mixed effects analyses, final model selection is based on measures of goodness of fit, such as the Akaike Information Criterion (AIC) or a refinement of the AIC corrected for sample size and penalized for the number of parameters in the model (AICC).[4,15] Another commonly used fit criterion is the [Schwarz] Bayesian Information Criterion (BIC or SBC). While important, choice of the optimal covariance structure is based on the smallest AICC or BIC, as well as other factors. While analogous fit indices may be calculated for GEE and GLMM, there is less consensus about their use in comparing covariance structures.

In addition to formal fit criteria, inspection of residuals is required to assess model fit. Models that assume normality, such as t-tests, ANOVA, linear regression and mixed effects require that residuals be independent and normally distributed. Statistical software can automatically produce diagnostic plots for residuals. Standardizing residuals is a convenient way of identifying residual values falling outside the range of −2 to 2 (approximately a 95% confidence band). As with the raw data, plotting the residuals against the expected distribution of residuals (quantile–quantile plot) is another effective method to assess normality, extreme values and model fit (see section 3.1). The residuals in a quantile–quantile plot should follow a diagonal line.

In addition to residual analysis, many statistical programs provide influence diagnostics.[16] These may give an indication of the stability of the model for this data, as well as help to identify extreme values. A commonly used methodology is to delete one case at a time from the analysis to determine the degree of influence (e.g., Cook's distance[17]) the observation exerts on the results. One is cautioned to remember that the results of diagnostic analysis depend on the model specifications. For example, an observation can be highly influential and/or an outlier because: (1) the data has an error (e.g., transcription), (2) there is measurement error (e.g., calibration of equipment), (3) the wrong distribution is specified (e.g., assumed normality when non-normally distributed) and (4) it is a valid observation that infrequently occurs. Options are to check the data and if no error can be found then to change the model by transforming the data, or select a different covariance structure or outcome distribution. Outliers can be the most important and noteworthy data points of rare occurrences as they can point to a model misspecification. The task is to develop a model that fits the data, not to develop a set of data that fits a particular model.

3. Examples (Mixed Effects Models)

3.1 Repeated measures — cross-sectional

In this example investigating aging and anti-fungal immunity (unpublished data with permission, Thekkiniath & Montgomery, 2015[18]), human polymorphonuclear cells from young (21–30 years old) and old (≥65 years old) adults were stimulated with C-type Lectin Receptor (CLR) ligands to study the expression of polymorphonuclear cell activation markers such as CD62L. The values are normalized — divided by their unstimulated (control) values — and thus represent the fold change from their control levels (Fig. 3).

While model fit indices may be compared quantitatively, frequently graphic measures are employed. Investigation of the optimal covariance structure in Fig. 4 shows unstructured to be the best selection for this model.

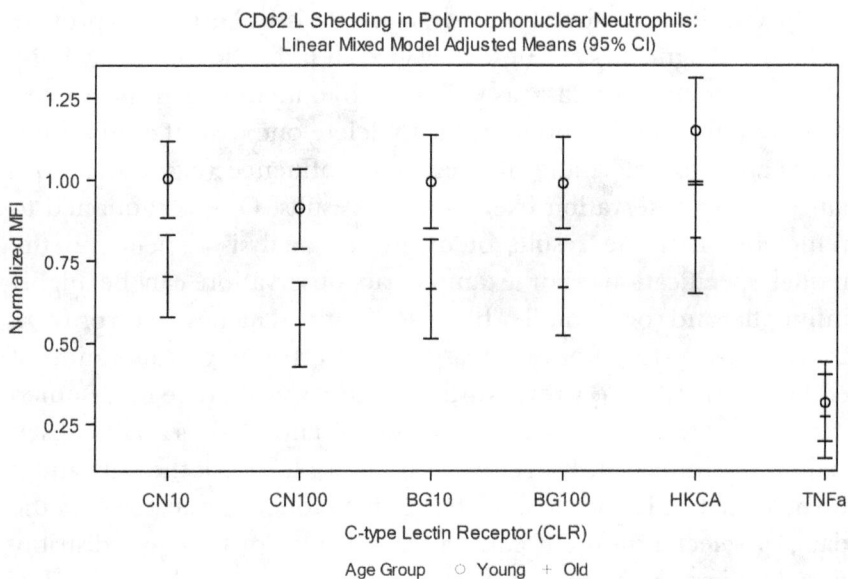

Fig. 3. Anti-fungal Immunity: Effect of aging on polymorphonuclear leukocyte activation in response to C-type Lectin Receptor (CLR) ligands. Normalized (fold change from control level) Mean Fluorescence Intensity (MFI), L-selectin CD62L fold change by C-type Receptor ligand stimulation: Means and 95% confidence intervals. CLR stimulation conditions: CN10, curdlan 10 μg/ml; CN100, curdlan 100 μg/ml; BG10, B-glucan 10 μg/ml; BG100, B-glucan 100 μg/ml; HKCA, heat-killed *C. albicans* 10^7 cells/ml; TNF*a*, Tumor Necrosis Factor α μg/ml. Means and 95% confidence limits from a linear mixed-effects model (adjusted for covariates: gender, race, number of comorbid conditions and number of current medications) depicted for young (21–30 years old) and older (≥65 years old) adults. Unpublished data with permission, Thekkiniath & Montgomery (2015).[18]

Continuing with the present example, Fig. 5 shows good model fit by residual examination. Studentized (standardized) residuals appear normally distributed (mean close to 0, standard deviation of 1); they are symmetric, with few outliers (largely ranging between −2 and 2) and the quantile-quantile plot approximates a straight line.

3.2. *Repeated measures — longitudinal (over time)*

In addition to previously described data visualizations, for repeated measurements over time spaghetti plots (individual level data

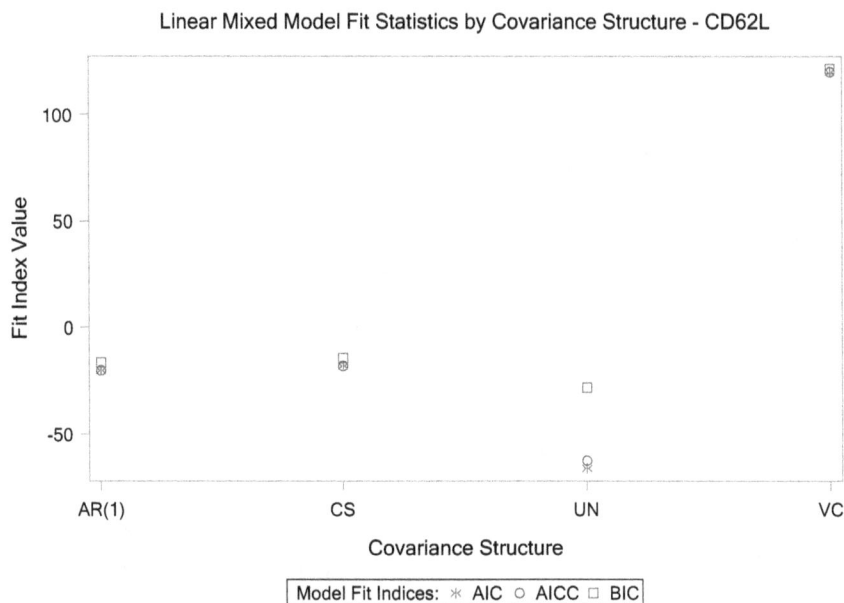

Linear Mixed Model Fit Statistics by Covariance Structure - CD62L

Fig. 4. Linear mixed-effect model fit indices for L-selectin, CD62L, under various C-type Lectin Receptor (CLR) ligand stimulations. Lower values indicate a better fit; thus, the unstructured covariance structure has the best fit based upon all three goodness of fit criteria. Note that unstructured covariance structure has a much lower AIC value than AICC as the former is not adjusted for the additional covariance parameters of the latter. Other structures have similar numbers of parameters. Unpublished data with permission, Thekkiniath & Montgomery (2015).[18]

connected by solid lines, Fig. 6) are frequently useful in gauging the variability of trajectories among individuals or across time periods. Spaghetti plots may provide insights to informing longitudinal models. For example, the trajectories over time may not be well characterized by a straight line but may require a single turning point (quadratic) or even two turning points (cubic) to enhance model fit and interpretation. Variation among subjects or within groups may become larger or smaller over time, which can be partially addressed by model specifications (see section 2.2 Covariance Structures). Patterns within the spaghetti plot may suggest the use of fixed or random effects in subsequent analyses (see topic Mixed Effects).

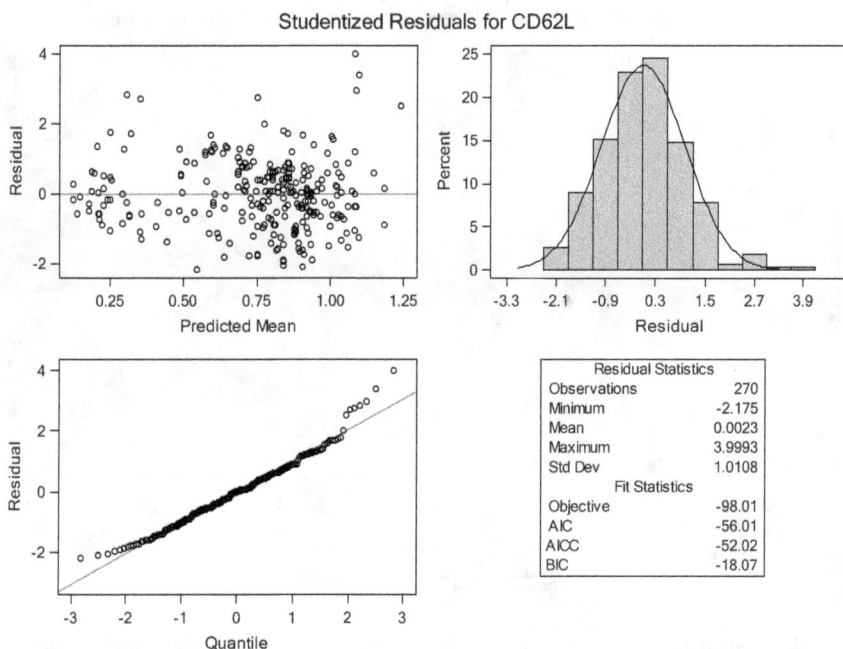

Fig. 5. Examination of Studentized residuals for L-selectin, CD62L, fold change under C-type Lectin Receptor (CLR) stimulation conditions (previous example) using a linear mixed-effects model employing an unstructured covariance structure. Unpublished data with permission, Thekkiniath & Montgomery (2015).[18]

We illustrate a longitudinal model with an investigation of CD86 expression over a 28-day period for inflammatory monocytes in two age groups of subjects after vaccination for influenza (adapted from Mohanty *et al.* (2015)).[19] Figure 6 is a spaghetti plot showing a subset of the raw data connected by solid lines; Whereas, Figure 7 shows the mean data for the two groups. One can see linear trends in the values over time, so we might expect that there is a relationship between measurements within an individual which might degrade over time (measurements closer to each other are more alike than those further apart).

As part of the determination of the best covariance structure for this data, the following plot was produced (Fig. 7), showing first-order autoregressive to have the best (lowest) among the model fit indices for a mixed model comparing CD86 expression in young and older

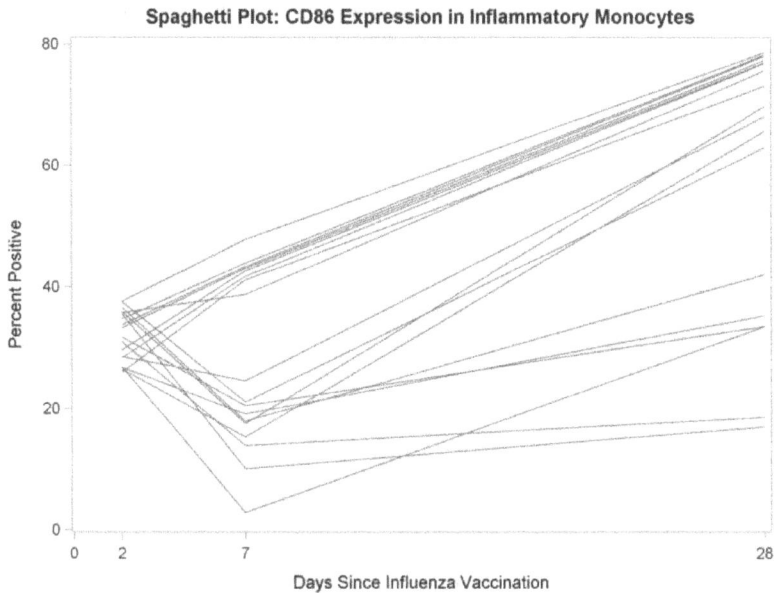

Fig. 6. Spaghetti plot: CD86 Expression in inflammatory monocytes over time since influenza vaccination. Subject-level data (subset) connected by lines for percent positive CD86 in inflammatory monocytes for both young (ages 21–30 years) and older (ages ≥65 years) subjects receiving the influenza vaccine. Time is number of days since vaccination. Data adapted with permission, Mohanty *et al.* (2014).[19]

adults over time. Figure 8 shows the final Model for the two groups over time.

4. Multiple Comparisons

4.1. *Many separate models versus a unified model*

Estimates derived from the same model share the same error distribution; thus, when a single confirmatory model is run *post-hoc* adjustments are not routinely undertaken. However, exploratory analyses involve many models and when estimates are derived from separate models then adjustments for multiple comparisons should be performed. When inference is applied to many model terms the situation of multiple comparisons arises; by chance one or more terms may be significant. The familywise error rate increases with the

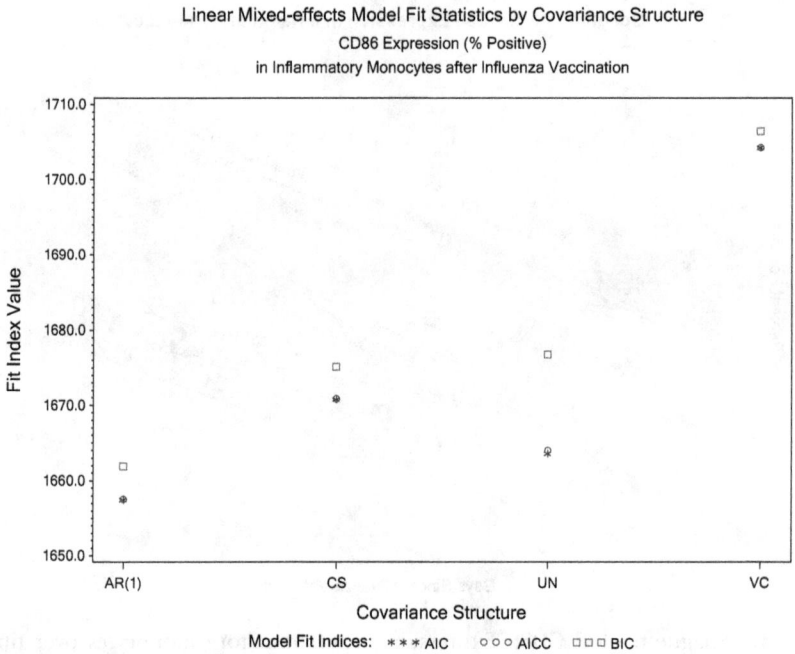

Fig. 7. Linear mixed model fit indices for CD86 percent expression in inflammatory monocytes over time for young (ages 21–30 years) and older (ages ≥65 years) age
groups after influenza vaccination. A first-order autoregressive covariance structure
was selected as the best fit. Data adapted with permission, Mohanty *et al.* (2015).[19]

number of tests performed.[20] For a Type I error of 0.05, given 100
tests on average 5 appear significant by chance. When multiple tests
are performed, the original Type I error is not maintained and the
probability of false positive results increases.

Multiple comparisons approaches include single-step methods and
sequentially rejective (stepwise) methods. With single-step methods,
where a single critical value is calculated against which tests are compared, directionality of the mean difference and confidence intervals can be constructed; however, they generally have lower power
(Type II error increases). Most sequentially rejective (stepwise)
methods test mean equality, not directionality and test null hypotheses against calculated significance levels. These methods are less
conservative than the corresponding single-step procedures.

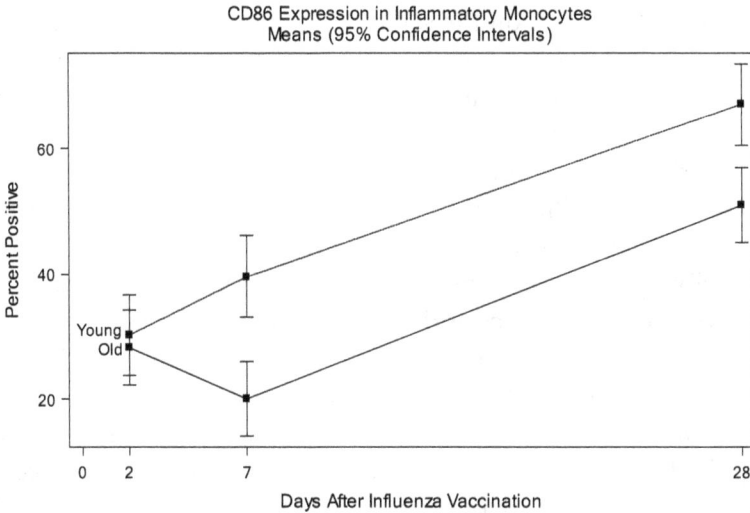

Fig. 8. Group-level data (means and 95% confidence limits) for percent positive CD86 expression over time for inflammatory monocytes in both young (ages 21–30 years) and older (ages ≥65 years) age groups after influenza vaccination. Data adapted with permission, Mohanty *et al.* (2015).[19]

Bonferroni and Sidak tests are single-step methods that adjust for the number of tests compared. The Bonferroni–Holm and Sidak–Holm methods[21] modify these to become "step down" methods — e.g., one adjusts the most significant term, then the second and so on to the least significant term.[22] The Hochberg procedure[23] is a "step up" procedure and works in the reverse direction; starting by adjusting the least significant term and stepping up to the most significant term. Both Holm's and Hochberg's methods were developed to maintain control over the family wise error rate. Most statistical programs provide options to have the significance level adjust for multiple comparisons.

5. Summary

Translational studies of human immunology require analytic models that account for heterogeneous samples, correlation among independent variables and among outcomes, control of covariates and repeated observations on the same subject either cross-sectionally or longitudinally. Inferences may be invalid if statistical assumptions are violated. Methods to visualize data distributions, select appropriate statistical tests and (where applicable) account for repeated measurements and multiple comparisons should be applied. When interpreting a set of probabilities (P-values) not corrected for repeated within subject correlation or multiple comparisons, readers must consider the possibility of a Type I error and overestimation of statistical significance. It is often best to consider these unadjusted models as exploratory, intended to generate hypothesis that can be tested with more focused experiments. Many of the concepts discussed in this chapter pertain widely to translational and biomedical research and should be applied at the design stage of a study, as they affect sample size calculations and analytic plans. It is our hope that these approaches become more commonplace to improve the analysis of experimental data in immunology.

Acknowledgements

Supported in part by the Yale Claude D. Pepper Older Americans Independence center (p30 AGO21342).

References

1. Medicare Current Beneficiary Survey. [https://www.cms.gov/Research-Statistics-Data-and-Systems/Research/MCBS/index.html?redirect=/MCBS/]. Centers for Medicare & Medicaid Services, 18 July 2013. Last accessed date, August 2015.
2. Sauler M, Leng L, Trentalange M *et al.* (2014) Macrophage migration inhibitory factor deficiency in chronic obstructive pulmonary disease. *Am J Physiol Lung Cell Mol Physiol.* **306**(6): L487–L496.
3. Hollander M, Wolfe DA. (1973) *Nonparametric Statistical Methods.* John Wiley & Sons, New York, NY.
4. Wolfinger RD. (1993) Covariance structure selection in general mixed models. *Comm Stat Stimul* **22**(4): 1079–1106.
5. van Duin D, Mohanty S, Thomas V *et al.* (2007) Age-associated defect in human Toll-like Receptor-1/2 function. *J Immunol* **178**(2): 970–975.

6. Panda A, Chen S, Shaw AC *et al.* (2013) Statistical approaches for analyzing immunologic data of repeated observations: A practical guide. *J Immunol Methods* **398–399**:19–26.

7. Searle SR, Casella G, McCulloch CE. (1992) *Variance Components.* John Wiley & Sons, Hoboken, NJ.

8. Zimmerman DL, Nunez-Anton V. (2001) Parametric modelling of growth curve data: An overview (with discussion). *Test* **10**:1–73.

9. Littell RC, Milliken GA, Stroup WW *et al.* (2006) *SAS® for Mixed Models,* 2nd ed, SAS Institute, Cary, NC.

10. Panda A, Qian F, Mohanty S *et al.* (2010) A generalized, age-associated defect in Toll-like Receptor function in primary human dendritic cells. *J Immunol* **184**(5):2518–2527.

11. Brown H, Prescott R. (1999) *Applied Mixed Models in Medicine.* John Wiley & Sons, New York, NY.

12. Collett D. (2003) *Modelling Binary Data,* 2nd ed. Chapman & Hall, London.

13. Hilbe JM. (2007) *Negative Binomial Regression.* Cambridge University Press, New York, NY.

14. Liang K-Y, Zeger SL. (1986) Longitudinal data analysis using generalized linear models. *Biometrika* **73**: 13–22.

15. Bozdogan H. (1987) Model selection and Akaike's Information Criteria (AIC): The general theory and its analytical extension. *Psychometrika* **52**: 345–370.

16. Martin RJ. (1992) Leverage, influence, and residuals in regression models when observations are correlated. *Commun Stat Theory.* **21**: 1183–1212.

17. Cook RD. (1977) Deletion of influential observations in linear regression. *Technometrics.* **19**:15–18.

18. Thekkiniath J, Montgomery R *et al.* (2015) Effect of aging on C-type lectin receptor expression and function in human neutrophils. Unpublished data with permission.

19. Mohanty S, Joshi SR, Ueda I *et al.* (2015) Prolonged pro-inflammatory cytokine production in monocytes is modulated by interleukin 10 after influenza vaccination in older adults. *J Infect Dis.* **211**: 1174–1184.

20. Shaffer JP. (1995) Multiple hypothesis testing. *Ann Rev Psychol* **46**: 561–584.

21. Holm S. (1979) A simple sequentially rejective multiple test procedure. *Scand J Stat* **6**:65–70.

22. Westfall PH, Tobias RD, Rom D *et al.* (1999) *Multiple Comparisons and Multiple Tests Using the SAS System.* SAS Institute, Cary, NC.
23. Hochberg Y, Benjamini Y. (1990) More powerful procedures for multiple significance testing. *Stat Med* **9**: 811–818.

Chapter 11

Systems Approaches to Autoimmune Diseases

Wan-Uk Kim[*,**,††], Sungyong You[†,‡], Daehee Hwang[§,¶]

[†]*Division of Cancer Biology and Therapeutics, Departments of Surgery and Biomedical Sciences, Samuel Oschin Comprehensive Cancer Institute, Cedars-Sinai Medical Center, Los Angeles, CA, 90048, USA*

[‡]*Sungyong.You@cshs.org*

[§]*Department of New Biology and Center for Plant Aging Research, Institute for Basic Science, DGIST, Daegu 711-873, Korea*

[¶]*dhwang@dgist.ac.kr*

[**]*Division of Rheumatology, Department of Internal Medicine The Catholic University of Korea, St. Mary's Hospital Seoul, 137-701 Korea*

[††]*wan725@catholic.ac.kr*

1. Introduction

Human autoimmune diseases arise from the complex interplay of innate and acquired immune systems. Molecular and cellular interactions underlying this interplay are highly complex. Different types

---

*Corresponding author.

of cells such as macrophages, dendritic cells, and *T* and *B* cells crosstalk via secretion of various cytokines, called cytokine networks.[1,2] Furthermore, in the individual cells, several signaling pathways such as toll-like receptors (TLR) and JAK-STAT signaling pathways are collectively or serially activated during disease progression and interact with each other. The complex interplay at both the molecular and cellular levels defines autoimmune disease phenotypes.[3–5] Thus, understanding the complex interplay at the molecular and cellular levels is crucial as they will allow us to develop new therapeutic strategies for the autoimmune diseases or to accurately determine the state of the autoimmune diseases for the optimization of therapeutic options.[6–8]

The molecular and cellular interplay can be viewed as a multilayered interaction structure, shown in Fig. 1, which includes five layers defined based on the complexity of the interplay. In the multilayered structure, immune cells are located at the center (3rd layer in Fig. 1). The lower layers include such immune receptors as TLRs,

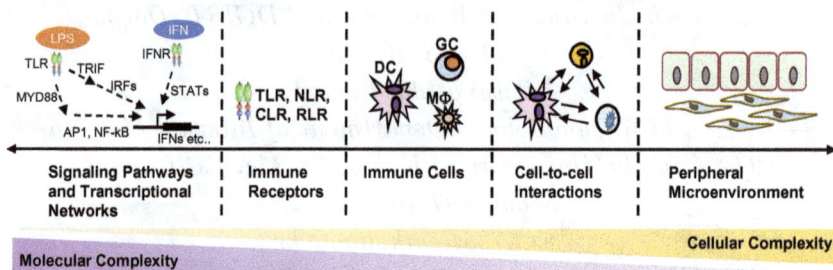

Fig. 1. Multilayered Complexity in Immune Systems. The complexity of immune systems can be abstracted as occurring in different level of layers based on molecular and cellular complexity. The basal unit of immune system is the cells, and zooming in on 'Immune cells' will reveal information about the immune receptors and signaling pathways and transcription factors that program immune cells to induce a particular immune response. Zooming out from 'Immune cells' will reveal the cellular interactions and peripheral microenvironment that influence the programming of the cells to generate a particular immune response. LPS, lipopolysaccharide; IFN, interferon; TLR, toll-like receptor; NLR, NOD-like receptor; CLR, C-type lectin receptor; RLR, RIG-I-like receptor; DC, dendritic cell; GC, granulocyte; MΦ, macrophage.

NOD-like receptors (NLRs), RIG-I-like receptor (RLRs), and inter-leukin and cytokine receptors (2nd layer), and their downstream signaling networks including NF-kB and JAK-STAT pathways (1st layer). These layers can provide more detailed information of the molecular interactions in individual cells. On the other hand, the 4th layer provides interactions between different types of cells, and the 5th layer further provides information on the influence of the multi-cellular interplay in tissue microenvironments. In this structure, molecular complexity increases toward the lower layers, while cellular complexity increases toward the higher layers. Thus, a unified understanding of cellular and molecular interplay can enhance understanding of the pathophysiological processes associated with the initiation and progression of autoimmune diseases.

Owing to the development of high-throughput technologies such as microarrays, next generation sequencing technologies, and mass spectrometry-based methods,[9-12] the abundance of molecules (genes, proteins, and metabolites) and their interactions have been profiled at the global level from various types of cells and tissues under autoimmune disease states.[13-17] Using these global measurements, systems approaches have been introduced to decode the multi-layered networks that can provide the molecular and cellular interplay. The resulting multi-layered networks can be used to identify 1) novel therapeutic targets to meet unmet clinical needs and 2) biomarkers to predict the prognosis and thereby to optimize therapeutic options. Herein, we summarize systems approaches that our group has developed for this purpose and their applications to rheumatoid arthritis (RA).

2. Decoding Complexity in Autoimmune Disease

2.1. *Molecular target discovery via network analysis*

Conventional approaches including linkage mapping or genome-wide association studies (GWAS) have been applied to identify the genes associated with the pathogenesis of autoimmune diseases.[18,19] For example, Begovich *et al.*[18] performed a linkage analysis of 840 individuals with RA from 463 families. They found an association with the minor allele (T) of a missense SNP (R620W [rs2476601, 1858C→T]) in the protein tyrosine phosphatase non-receptor type 22 gene (PTPN22). Also, Hom et al.[19] performed a GWAS study for systemic lupus erythematosus (SLE) and found that the genes encoding B lymphoid tyrosine kinase (BLK), C8orf13 (chromosome 8p23.1), integrin alpha M (ITGAM), and integrin alpha X (ITGAX) were associated with the pathogenesis of SLE. While these approaches provide a list of potential genes that are associated with the pathogenesis of autoimmune diseases, they involve no analysis of the interactions of these genes to understand collective functions of these genes and alterations of their functions in autoimmune diseases. With no understanding of the collective functions of the potential genes via interactions between them, these approaches are limited to reliably identifying key genes or pathways associated with disease pathogenesis.

To grasp the interactions among the potential genes, systems approaches have been introduced to identify disease-associated genes from gene or protein expression profiling of the tissues or cells collected from autoimmune diseases and then to develop network models that describe the interactions between the disease-associated genes. Recently, enormous amounts of global data (e.g. gene or proteomic expression profiles) of immune cells and tissues collected from various autoimmune diseases, such as RA, SLE, dermatomyositis, polymyositis, and systemic scleroderma, have been deposited to public databases.[8] Using these global datasets, network-based methods have been applied to predict key molecules (e.g. genes or proteins) indicating 1) novel biomarkers that can be used to monitor the state of disease initiation and progression or the response to the therapy and 2) new therapeutic targets for the diseases.[20,21]

For instance, our group previously introduced a systems approach to reconstruct RA networks using 3 gene expression datasets obtained from synovial tissues.[22] This study found that 19 key transcription factors (TFs) can be responsible for 55% of the alterations of gene expression caused by RA. Among them, we choose NFAT5, a osmoprotective transcription factor that can cause the alteration of key modules associated with cell cycle, cell death/survival, antigen presentation, T cell activation, and angiogenic factors in the RA network.[23,24] We also reconstructed a molecular network that describes such fibroblast-like synoviocytes (FLS) dominant cellular functions as cell migration and invasion compared to synovial macrophages (SM) through gene expression profiling of FLS and SM under unstimulated and IL-1β-stimulated conditions. RA-FLS migrate and attach to cartilage and bone and then destroy them by secreting proteases within the RA joints.[25,26] Thus, decoding the mechanisms underlying the migration and invasion of RA-FLS is essential to the understanding of RA pathogenesis. For this reason, we first identified functional network modules from the network (Fig. 2a). We then quantitatively assessed the contribution of the modules to the perturbation of the network based on differentially expressed genes (DEGs) from the comparison of RA-FLS versus OA-FLS stimulated without or with IL-1β. This analysis revealed a list of key functional modules, including FLS-dominant modules (EMT, ECM, and WNT) and newly acquired pro-inflammatory modules (IC, AG, RCP, JS, CP, and CC) newly acquired by IL-1β (Fig. 2a).[27] We next selected 13 and 15 genes with regulator activities (e.g. transcriptional regulators and/or signaling molecules) from the genes involved in FLS-dominant modules and pro-inflammatory modules, respectively. In particular, the *in vitro* and *in vivo* functional assays revealed that TWIST1 and POSTN are crucial for the migration and invasion of FLS stimulated with IL-1β (Fig. 2b). In the meantime, IFNAR1/2 and IRF9 were identified from 15 regulators responsible for pro-inflammatory immune responses such as cytokine production in RA-FLS upon IL-1β stimulation (Fig. 2c). This result suggested that interferon signaling pathways can be activated by IL-1β stimulation in RA-FLS, leading the promotion of IFN-responsive gene transcription possibly via IRF9.

In an analogous manner, Chaussabel *et al.* have developed a modular framework for the purpose of analyzing blood transcriptome in SLE patients.[28] This modular framework was used to visualize differential regulation of IFN-responsive genes at the individual level in other studies,[29,30] which concluded that the expression of IFN-responsive genes is a stable characteristic of SLE patients and may reflect an active state of the IFN pathway in SLE.[29,30] Collectively, these network-based methods followed by global molecular profiling can serve as a useful strategy for unbiased discovery of potential molecular targets and biomarkers.

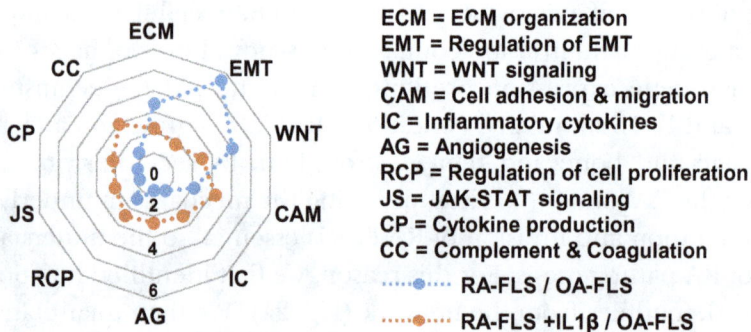

ECM = ECM organization
EMT = Regulation of EMT
WNT = WNT signaling
CAM = Cell adhesion & migration
IC = Inflammatory cytokines
AG = Angiogenesis
RCP = Regulation of cell proliferation
JS = JAK-STAT signaling
CP = Cytokine production
CC = Complement & Coagulation

⚬⋯⋯ RA-FLS / OA-FLS
⚬⋯⋯ RA-FLS +IL1β / OA-FLS

(a)

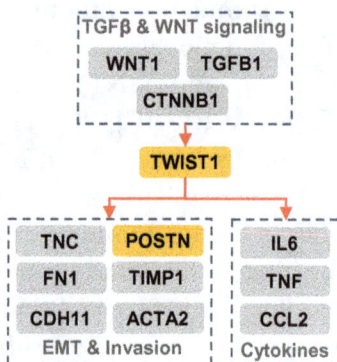

(b)

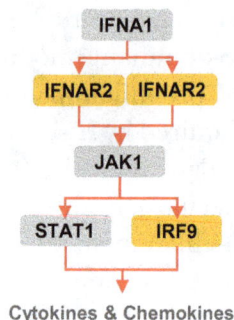

(c)

Fig. 2. Modular Abstraction Facilitate Extraction of pathological implication. (a) Modular enrichment scores (MES) computed from the FLS network model (See details in Ref. 22). Each line on the radar plot represents a MES score (from 0 to 6). RA-FLS versus OA-FLS (blue). RA-FLS +IL1β versus OA-FLS (orange). (b) Subnetwork embedded in RA-FLS. (c) Newly acquired subnetwork in response to IL1β stimulation in RA-FLS.

2.2 *Extracting cell-specific information from heterogeneous samples*

As mentioned earlier, understanding the roles of individual cell types in the pathogenesis of autoimmune diseases is essential to identify key molecules associated with particular pathophysiology in disease tissue microenvironments or particular cellular processes underlying the responses to therapeutic agents. However, the previous systems approaches have focused primarily on the single-layer network models representing the ensemble view of functions of multiple types of cells, rather than the multi-layered network models that provide cell-specific functions and interactions among different types of cells as shown in the 1st to 3rd layers in Fig. 1. Several computational methodologies have tackled this issue to extract cell type-specific information from the global data of the tissues consisting of different types and proportions of cells.[31–34] Here, we thus utilized the computational methodologies to extract cell-specific information and then demonstrated that the cell-centered view can provide increased resolution and interpretability of the mechanisms underlying disease pathogenesis as well as the disease status from the clinical samples.

Biologic agents, including TNF-α inhibitors, IL-6 blockers, and B-cell ablating agents, have been widely used in the treatment of autoimmune diseases such as RA, SLE, and other inflammatory diseases.[35–38] While these agents facilitate efficacious therapies for many of inflammatory diseases, the mechanisms underlying their efficacy are still unclear. Indeed, 30–40% of RA patients treated with anti-TNF agents failed to achieve complete remission of RA.[39, 40] Besides the anti-TNF agents, patients treated with other biologic agents have faced similar situations.[41–43] Hence, identification of differential roles of the participating cells in disease pathogenesis has been a key question in developing therapeutic interventions or a guideline for selecting a desired therapeutic strategy.

In order to gauge the usefulness of cell-centered information, we attempted to extract the information for differential roles of cell types before and after treatments of three agents including infliximab (IFX; anti-TNF-α), tocilizumab (TCZ; anti-IL-6), and rituximab (RTX; anti-CD20) from three mRNA expression datasets from synovial tissues of RA patients.[44–46] The information was extracted based on the cell type enrichment analysis (CTen) software.[31] Figure. 3a shows that 8 out of 10 cell types were highly affected by the agents, supporting why they are effective mostly for RA patients. As expected, four innate immune-related cell types (dendritic cells, monocytes,

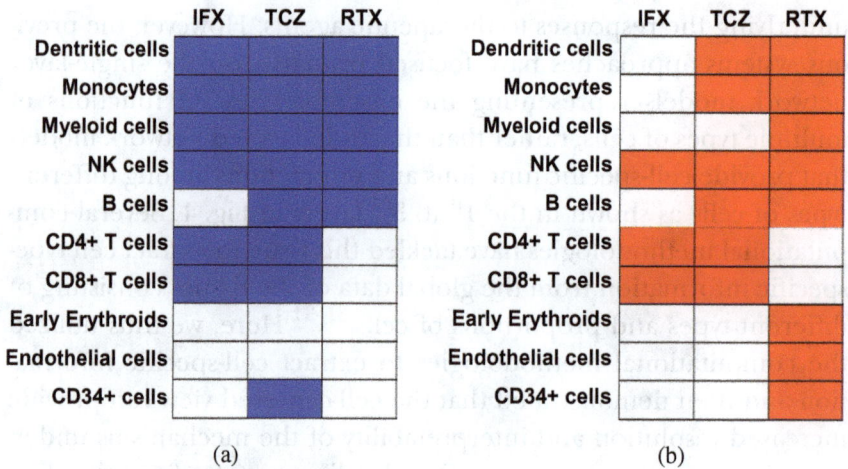

Fig. 3 Impact of Major Cytokine Interventions in the Context of Cellular Activity. Using the mRNA expression data from synovial tissues of RA patients treated with those biologics, three sets of down-regulated genes were used to compute cell type specific enrichment scores using Cten software (http://www.influenza-x.org/~jshoemaker/cten/). (a) Repressed cell subsets by treatment of IFX, TCZ, or RTX. Blue indicates significant enrichment ($-\log_{10}$ Benjamini-Hochberg adjusted $P > 2$, which is corresponding to $P < 0.01$) of cell subsets. White is not significant. (b) Activated cell subsets in poor responders compared to good responders with respect to each agent. Red is significantly enriched and white is not significant.

myeloid cells, and natural killer cells) were significantly affected by the treatment with all the three agents (Fig. 3a). Of note, the adaptive immune cells, including T cells, were reduced by treatment with infliximab or tocilizumab, suggesting that the elevated TNF-α and IL-6 may be necessary for adaptive immune activity. However, B-cell-related transcriptional changes were not affected by infliximab therapy (Fig. 3a), suggesting that B-cell-targeted therapy may be effective for anti-TNF-α resistant cases. Indeed, tocilizumab and rituximab have been approved for the treatment of RA patients who are refractory to TNF-α inhibitors.[47,48] In addition, the tocilizumab therapy represses CD34+ cells. In fact, this cell type is known to be involved in the generation of endothelial cells in RA conditions, leading to neovascularization in the RA synovium.[49] This suggests that one of therapeutic roles of tocilizumab might be repression of neovascularization through suppression of differentiation capacities into endothelial cells. Collectively, these data indicates that the approaches for understanding cell-centered information can potentially be a useful strategy to understand modes of action for treatments with therapeutic agents to diverse subgroups of patients with autoimmune diseases.

In addition to the evaluation of the impact of therapeutic intervention, predicting a patient's responsiveness to a particular agent prior to treatment of patients is extremely important to increase the efficacy of therapy and to prevent side effects. The efficacy of therapeutic agents targeting a molecule expressing in particular cell types, such as CD20+ B-cells, is supposed to be linked with the state or proportion of the cell type under disease conditions. Therefore, the state of particular cell types at the time of initial treatments might be a good measure for estimating the efficacy for the therapeutic agent. Along this line, we attempted to estimate the relative state of cell types before the treatment in RA patients with poor responses to each agent, compared to those with good responses (Fig. 3b). As a result, for patients with high inflammatory activity of *T* cells and NK cells, the prognosis to treatment with infliximab or tocilizumab is poor after the treatment. Also, the patients with good responses to the tocilizumab therapy might be additionally dependent on the activities of dendritic cells, monocytes, and *B* cells at the initial stage of treatment. Unlike infliximab and tocilizumab, the responsiveness to rituximab therapy is not associated with *T* cell activity, but is sensitive to the activity of other cell types, including CD34+, early erythroid, and endothelial cells. This estimate of cell type state is consistent with the results of previously studies on patients' responsiveness to the treatments. For example, the baseline activity of dendritic cells and monocytes is more active in RA patients with poor responses to the tocilizumab therapy, compared to those with good responses.[45,50,51] Furthermore, the poor responders to the rituximab therapy revealed the detection of highly active angiogenesis in synovium.[52] Given the desire to investigate their differences in the responses to therapeutic agents, we anticipate that the approaches for understanding cell-centered information can offer better prognostic values than indications from individual molecules alone.

3. Challenges and Perspectives

Several important insights have emerged from studies focusing on a single layer (e.g. signaling pathways, or cell-to-cell interactions), but such insights do not offer a global picture of autoimmune disease phenotypes. Most autoimmune diseases originate from complex interplay at the molecular and cellular levels. Therefore, future research should seek an integrated understanding of molecular and cellular interplay through decoding multi-layered networks in Fig. 1. This analysis of the multi-layered networks allows us to develop a unified molecular and cellular stratification that can improve identification of a group of patients who benefit from treatment with a therapeutic agent.

With this line, a proposed research strategy can be configured to two phases, exploratory and translational. The exploratory phase is to measure the global-scale dynamic change of transcripts or proteins under clinical conditions (Fig. 4a). This is followed by the reconstruction of molecular networks based on the co-expression of regulators and their putative targets and/or on the enrichment of cis-elements in these targets, as well as the extraction of cell state information based on computational deconvolution of heterogeneous samples into cell types (Fig. 4b). Using the information from the molecular networks and the cell-centered view, we can identify cell types implying the potential molecular regulators-target relationship that can modulate cell state in clinical conditions. Some of the predictions of the cell types implying the regulator-target relationship obtained from the exploratory phase, which are considered pathogenic drivers leading to inflammatory disease phenotypes, are experimentally tested by the perturbation of predicted TFs and/or signaling molecules using siRNA and siRNA or stimulation with drugs targeting the predicted regulators in the cell line models (Fig. 4c). Then, we measure downstream molecular profiles of the

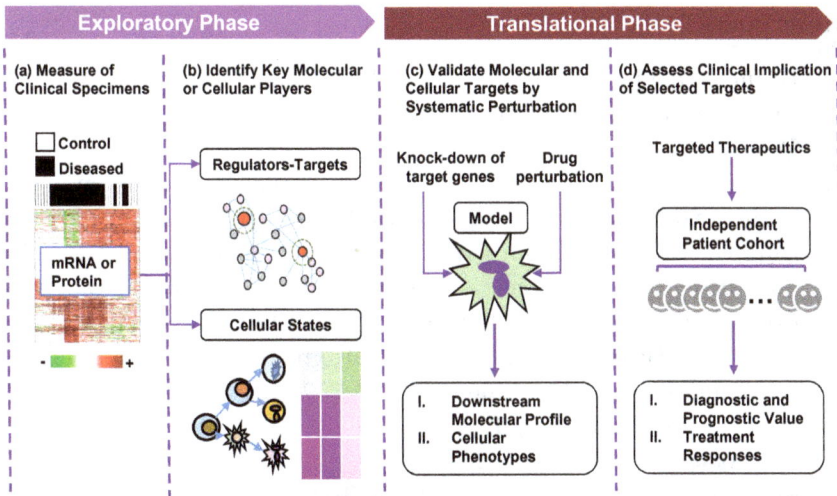

Fig. 4 A proposed Strategy for Systems Approach to Autoimmune Diseases. The diagram shows the four major steps in translational systems approaches to complex autoimmune diseases. (a) Measure of clinical specimens. (b) Identify key Molecular or cellular players. (c) validate Molecular and cellular targets by systematic perturbation. (d) Assess clinical implication of selected targets.

regulators by perturbations and stimulation with drugs, such as increase or decrease of cytokine secretion or cell proliferation and migration. Eventually, the predicted models of selected cell types and regulators are then updated based on the experimental data. In second step of the translational phase (Fig. 4d), the most attractive regulator as a therapeutic target is tested using the cells isolated from patient blood or disease-affected tissues and studied *ex vivo* using global profiling approaches as described in Fig. 4a. If alterations in molecular networks and cellular behaviors can be associated with particular autoimmune disease phenotypes, then the molecular networks and cell-centered information will be helpful in the diagnosis and prognosis of autoimmune disorders and in the selection of therapeutic interventions. Given the unmet needs in the diagnoses, monitoring, and treatment of autoimmune diseases, this proposed systems approach is clearly a highly compelling means to gaining insight into disease mechanisms and discovering therapeutic targets and novel biomarkers.

References

1. Deane KD, El-Gabalawy H. (2014) Pathogenesis and prevention of rheumatic disease: focus on preclinical RA and SLE. *Nat Revs Rheumatol* **10**: 212–228.

2. Brennan FM, McInnes IB. (2008) Evidence that cytokines play a role in rheumatoid arthritis. *J Clin Invest* **118**:3537–3545.

3. Ding C, Cai Y, Marroquin J *et al.* (2009) Plasmacytoid dendritic cells regulate autoreactive B cell activation via soluble factors and in a cell-to-cell contact manner. *J Immunol* **183**: 7140–7149.

4. Ganguly D, Haak S, Sisirak V *et al.* (2013) The role of dendritic cells in autoimmunity. *Nat Rev Immunol* **13**: 566–577.

5. Hurst J, von Landenberg P. (2008) Toll-like receptors and autoimmunity. *Autoimmun Rev* **7**: 204–208.

6. Kim WU, Sreih A, Bucala R. (2009) Toll-like receptors in systemic lupus erythematosus; prospects for therapeutic intervention. *Autoimmun Rev* **8**: 204–208.

7. Burmester GR, Feist E, Dorner T. (2014) Emerging cell and cytokine targets in rheumatoid arthritis. *Nat Revs Rheumatol* **10**: 77–88.

8. Pascual V, Chaussabel D, Banchereau J. (2010) A genomic approach to human autoimmune diseases. *Ann Rev Immunol* **28**: 535–571.

9. Sabido E, Selevsek N, Aebersold R. (2012) Mass spectrometry-based proteomics for systems biology. *Curr Opin Biotechnol* **23**: 591–597.

10. Price JV, Tangsombatvisit S, Xu G *et al.* (2012) On silico peptide microarrays for high-resolution mapping of antibody epitopes and diverse protein-protein interactions. *Nat Med* **18**: 1434–1440.

11. Soon WW, Hariharan M, Snyder MP. (2013) High-throughput sequencing for biology and medicine. *Mol Systems Biol* **9**: 640.

12. Patti GJ, Yanes O, Siuzdak G. (2012) Innovation: Metabolomics: The apogee of the omics trilogy. *Nature Rev. Mol Cell Biol* **13**: 263–269.

13. Zhernakova A, Withoff S, Wijmenga C. (2013) Clinical implications of shared genetics and pathogenesis in autoimmune diseases. *Nat Rev Endocrinol* **9**: 646–659.

14. Eizirik DL, Sammeth M, Bouckenooghe T *et al.* (2012) The human pancreatic islet transcriptome: Expression of candidate genes for type 1 diabetes and the impact of pro-inflammatory cytokines. *PLoS genetics* **8**: e1002552.

15. Serrano-Fernandez P, Moller S, Goertsches R *et al.* (2010) Time course transcriptomics of IFNB1b drug therapy in multiple sclerosis. *Autoimmunity* **43**: 172–178.

16. Jabbari A, Suarez-Farinas M, Dewell S *et al.* (2012) Transcriptional profiling of psoriasis using RNA-seq reveals previously unidentified differentially expressed genes. *J Invest Dermatol* **132**: 246–249.

17. Gervin K, Vigeland MD, Mattingsdal M *et al.* (2012) DNA methylation and gene expression changes in monozygotic twins discordant for psoriasis: Identification of epigenetically dysregulated genes. *PLoS genetics* **8**: e1002454.

18. Begovich AB, Carlton VE Honigberg LA *et al.* (2004) A missense single-nucleotide polymorphism in a gene encoding a protein tyrosine phosphatase (PTPN22) is associated with rheumatoid arthritis. *Am J Hum Genet* **75**: 330–337.

19. Hom G, Graham RR, Modrek B *et al.* (2008) Association of systemic lupus erythematosus with C8orf13-BLK and ITGAM-ITGAX. *N Engl J Med* **358**: 900–909.

20. Chiche L, Jourde-Chiche N, Pascual V *et al.* (2013) Current perspectives on systems immunology approaches to rheumatic diseases. *Arthritis Rheum* **65**: 1407–1417.

21. Sirota M, Butte AJ. (2011) The role of bioinformatics in studying rheumatic and autoimmune disorders. *Nat Rev Rheumatol* **7**: 489–494.

22. You S, Cho CS, Lee I *et al.* (2012) A systems approach to rheumatoid arthritis. *PLoS one* **7**: e51508.

23. Yoon HJ, You S, Yoo SA *et al.* (2011) NF-AT5 is a critical regulator of inflammatory arthritis. *Arthritis Rheum* **63**: 1843–1852.

24. Wu G, Zhu L, Dent JE *et al.* (2010) A comprehensive molecular interaction map for rheumatoid arthritis. *PLoS one* **5**: e10137.

25. Bartok B, Firestein GS. (2010) Fibroblast-like synoviocytes: Key effector cells in rheumatoid arthritis. *Immunol Revs* **233**: 233–255.

26. Lefevre S, Knedla A, Tennie C *et al.* (2009) Synovial fibroblasts spread rheumatoid arthritis to unaffected joints. *Nat Med* **15**: 1414–1420.

27. You S, Yoo SA, Choi S *et al.* (2014) Identification of key regulators for the migration and invasion of rheumatoid synoviocytes through a systems approach. *Proc Natl Acad Sci USA* **111**: 550–555.

28. Chaussabel D, Quinn C, Shen J *et al.* (2008) A modular analysis framework for blood genomics studies: Application to systemic lupus erythematosus. *Immunity* **29**: 150–164.

29. Landolt-Marticorena C, Bonventi G, Lubovich A *et al.* (2009) Lack of association between the interferon-alpha signature and longitudinal changes in disease activity in systemic lupus erythematosus. *Ann Rheum Dis* **68**: 1440–1446.

30. Petri M, Singh S, Tesfasyone H *et al.* (2009) Longitudinal expression of type I interferon responsive genes in systemic lupus erythematosus. *Lupus* **18**: 980–989.

31. Shoemaker JE, Lopes TJ, Ghosh S *et al.* (2012) CTen: A web-based platform for identifying enriched cell types from heterogeneous microarray data. *BMC Genom* **13**: 460.

32. Abbas AR, Wolslegel K, Seshasayee D *et al.* (2009) Deconvolution of blood microarray data identifies cellular activation patterns in systemic lupus erythematosus. *PloS one* **4**: e6098.

33. Shen-Orr SS, Tibshirani R, Khatri P *et al.* (2010) Cell type-specific gene expression differences in complex tissues. *Nat Methods* **7**: 287–289.

34. Gaujoux R, Seoighe C. (2013) CellMix: A comprehensive toolbox for gene expression deconvolution. *Bioinformatics* **29**: 2211–2212.

35. Choy EH, Isenberg DA, Garrood T *et al.* (2002) Therapeutic benefit of blocking interleukin-6 activity with an anti-interleukin-6 receptor monoclonal antibody in rheumatoid arthritis: A randomized, double-blind, placebo-controlled, dose-escalation trial. *Arthritis Rheum* **46**: 3143–3150.

36. Nishimoto N, Kishimoto T. (2006) Interleukin 6: From bench to bedside. *Nat Clin Practice Rheumatol* **2**: 619–626.

37. Edwards JC, Szczepanski L, Szechinski J *et al.* (2004) Efficacy of B-cell-targeted therapy with rituximab in patients with rheumatoid arthritis. *N Engl J Med* **350**: 2572–2581.

38. Jones RB, Tervaert JW, Hauser T *et al.* (2010) Rituximab versus cyclophosphamide in ANCA-associated renal vasculitis. *N Engl J Med* **363**: 211–220.

39. Lipsky PE, van der Heijde DM, St Clair EW *et al.* (2000) Infliximab and methotrexate in the treatment of rheumatoid arthritis. Anti-Tumor Necrosis Factor Trial in Rheumatoid Arthritis with Concomitant Therapy Study Group. *N Engl J Med* **343**: 1594–1602.

40. Weinblatt ME, Keystone EC, Furst DE *et al.* (2003) Adalimumab, a fully human anti-tumor necrosis factor alpha monoclonal antibody, for the treatment of rheumatoid arthritis in patients taking concomitant methotrexate: The ARMADA trial. *Arthritis Rheum* **48**: 35–45.

41. Buch MH, Bingham SJ, Seto Y *et al.* (2004) Lack of response to anakinra in rheumatoid arthritis following failure of tumor necrosis factor alpha blockade. *Arthritis Rheum* **50**: 725–728.

42. Yamanaka H, Tanaka Y, Inoue E *et al.* (2011) Efficacy and tolerability of tocilizumab in rheumatoid arthritis patients seen in daily clinical

practice in Japan: Results from a retrospective study (REACTION study). *Mod Rheumatol* **21**: 122–133.

43. Ramos-Casals M, Soto MJ, Cuadrado MJ *et al.* (2009) Rituximab in systemic lupus erythematosus: A systematic review of off-label use in 188 cases. *Lupus* **18**: 767–776.

44. Lindberg J, af Klint E, Catrina AI *et al.* (2006) Effect of infliximab on mRNA expression profiles in synovial tissue of rheumatoid arthritis patients. *Arthritis Res Ther* **8**: R179.

45. Ducreux J, Durez P, Galant C *et al.* (2014) Global molecular effects of tocilizumab therapy in rheumatoid arthritis synovium. *Arthritis Rheumatol* **66**:15–23.

46. Gutierrez-Roelens I, Galant C, Theate I *et al.* (2011) Rituximab treatment induces the expression of genes involved in healing processes in the rheumatoid arthritis synovium. *Arthritis Rheum* **63**: 1246–1254.

47. Suzuki T, Horikoshi M, Sugihara M *et al.* (2013) Therapeutic efficacy of tocilizumab in patients with rheumatoid arthritis refractory to anti-tumor-necrosis-factor inhibitors: 1 year follow-up with low-field extremity MRI. *Mod Rheumatol* **23**:782–787.

48. Kremer JM, Tony HP, TakPP *et al.* (2006) Efficacy of rituximab in active RA patients with an inadequate response to one or more TNF inhibitors. *Arthritis* Rheum **54**: S247–S248.

49. Hirohata S, Yanagida T, Nampei A *et al.* (2004) Enhanced generation of endothelial cells from CD34+ cells of the bone marrow in rheumatoid arthritis: Possible role in synovial neovascularization. *Arthritis Rheum* **50**: 3888–3896.

50. Richez C, Barnetche T, Khoryati L *et al.* (2012) Tocilizumab treatment decreases circulating myeloid dendritic cells and monocytes, 2 components of the myeloid lineage. *J Rheumatol* **39**: 1192–1197.

51. Sanayama Y, Ikeda K, Saito Y *et al.* (2014) Prediction of therapeutic responses to tocilizumab in patients with rheumatoid arthritis: Biomarkers identified by analysis of gene expression in peripheral blood mononuclear cells using genome-wide DNA microarray. *Arthritis Rheumatol* **66**: 1421–1431.

52. Gonzalez-Juanatey C, LlorcaJ, Vazquez-Rodriguez TR *et al.* (2008) Short-term improvement of endothelial function in rituximab-treated rheumatoid arthritis patients refractory to tumor necrosis factor alpha blocker therapy. *Arthritis Rheum* **59**:1821–1824.

Chapter 12

Genetic Mapping of Human Immune System Function

Arpita Singh[*,†] *and Chris Cotsapas*[†,‡,§,¶,**]

Department of Laboratory Medicine, Yale School
of Medicine, New Haven, CT USA
[†]*Department of Neurology, Yale School of Medicine*
New Haven, CT USA
[‡]*Department of Genetics, Yale School of Medicine*
New Haven, CT USA
[§]*Program in Medical and Population Genetics*
Broad Institute of MIT and Harvard, Boston, MA USA
[¶]*chris.cotsapas@yale.edu*

1. Introduction

Many aspects of human immune system function vary across individuals: This includes functional aspects such as relative sizes of immune cell subpopulations, or response to antigen challenge; and susceptibility to autoimmune, inflammatory, and infectious diseases. Whilst part of this variation can be attributed to environmental factors, a large proportion is heritable: This *heritability* (see glossary) is driven by genetic variation in various loci in the genome,

---

[¶]Corresponding author.

which must affect genes relevant to the underlying biology of the *trait*. Thus, by locating these variants *genetic mapping*, we can identify these genes and understand the underlying mechanisms of disease. In this chapter, we provide an overview of recent progress in genetic mapping across immune-related traits. While detailed discussion of each trait is beyond the scope of this chapter, we will first outline how genetic mapping is performed and then discuss specific examples that demonstrate its ultimate biological utility.

Genetic traits can broadly be divided into two classes of *genetic architecture*: (1) *monogenic* traits, where a single DNA *mutation* is causal; and (2) *polygenic*, where many *variants* contribute to disease in an additive manner but none are necessary and sufficient.[1] Monogenic disorders are generally rare and either follow a simple inheritance pattern in affected families by segregating according to Mendel's laws, or are caused by *de novo* mutations that arise during *meiosis* in one parent and are transmitted to affected offspring. Family-based studies have been successful in identifying the mutations in monogenic disorders through *linkage analysis* of large pedigrees for inherited mutations. More recently, large-scale DNA sequencing technologies have become cost-effective and have been applied to extended pedigrees or parent-affected offspring trios to identify *de novo* mutations.

Polygenic or *complex traits* do not exhibit clear patterns of inheritance, although there is obvious enrichment in families.[1] The *common disease, common variant* hypothesis predicts that the variants that drive risk in common, complex traits will themselves be common.[2] The sequencing of the human genome and subsequent exhaustive cataloging of variation in diverse populations has allowed the large-scale *genotyping* of common variation genome-wide.[3,4] Family-based studies have been largely unsuccessful at identifying complex trait loci. We now know this is due to the modest amount of risk encoded by each risk variant: The number of families required for *statistical power* of such experiment becomes impossible to collect.[5] In contrast, population-based *genome-wide association studies* (GWAS) where unrelated cases and controls are compared has been very fruitful, yielding tens of thousands of replicable associations to hundreds of traits covering virtually every human organ system and biological process.[6]

2. Monogenic Diseases of the Immune System

There are over 160 rare, monogenic disorders affecting the immune system. These include combined B and T cell immunodeficiencies, defects in antibody production, the complement cascade and insufficient function of cellular apoptosis, phagocytosis, ligand/receptor binding-mediated signaling events, DNA breakage and modification syndromes, as well as periodic fever syndromes.[7] Mapping the genes underlying these diseases has revealed not only the functions of those genes, but in several cases led to new insights on the function of the immune system and to the development of novel treatments.[8]

Insights from Association Studies in Inflammatory and Autoimmune Diseases

Over the last 15 years, much progress has been made in mapping the determinants of polygenic immune traits, with autoimmune and inflammatory diseases being the most studied (see Table 1 for a summary of discoveries across traits). Tens to hundreds of loci have been identified for susceptibility to several diseases, including inflammatory bowel disease (IBD[9]), multiple sclerosis (MS[10]), psoriasis,[11] rheumatoid arthritis (RA[12]), systemic lupus erythematosus (SLE[13]), and Type 1 diabetes (T1D[14]). From these results (and comparable ones in traits affecting other physiological systems), several general insights have emerged[6]:

1. *The common disease, common variant hypothesis is correct.* There are hundreds of statistically significant associations of common variants to disease risk, which have been replicated in independent samples and, often, in independent populations.
2. *The polygenic model is also correct.* Each trait has many tens of associations to independent loci, with projections suggesting as many as 1000 independent genetic effects in multiple sclerosis, likely a representative disease. This naturally implies that each risk variant accounts for a very small proportion of risk and so prediction of disease likelihood from genotype in healthy individuals is not likely to be useful.

Table 1.   Summary of GWAS Findings across Immune-Related Diseases and Traits. We list the number of independent loci reported in the literature as being significantly associated to each trait in each major continental population. Significance in GWAS is taken as $p < 5 \times 10^{-8}$, which corrects for the number of independent markers across the genome (see Ref. 6). Other authors report less stringent associations in their studies, which are considered suggestive and not included here. A real-time listing of all GWAS associations in the literature can be found at the NHGRI's GWAS catalog at www.genome.gov/gwastudies.

	Population	Samples	Loci	References
Autoimmune/ inflammatory Diseases				
MS	European	29,000	97	10
IBD	European	75,000	163	9
	Asian	1727	5	43
SLE	European	8000	30	44
	Asian	22,215	36	45
RA	European	100,000	101	12
	Asian	28,159	28	46
Celiac disease	European	25,000	40	47
T1D	European	20,000	40	14
Psoriasis	European	7,000	15	11
Atopic dermatitis	European	10,000	10	48
	Asian	7,964	18	49
Allergies (cat, dust-mite, pollen)		54,000	16	50
Allergic sensitization		15,000	10	51
IgA deficinecy	European	772	5	52
Infectious Diseases				
HIV and AIDS	European	9,630	9	36
	African American	1,748	2	36
Hepatitis C	European	1,362	1	33
Hepatitis B	Asian	6,387	22,930	
Dengue	Asian	8,697	2	53
Malaria	African	5,900	2	54
Tuberculosis	African	11,425	25,556	
Leprosy	Asian	11,140	6	57
Infectious leishmaniasis	Indian and Brazilian	2,959	2	28
Vaccine responsiveness				
Smallpox	Multiple	1720	22	39
Hepatitis B	Asian	5050	63,233	

3. *Most risk variants affect gene regulation rather than protein structure.*
    One of the puzzling observations from GWAS results has been
    that most association peaks do not overlap gene-coding regions,
    and when they do, the most associated markers fall outside tran-
    script boundaries. Two strands of evidence suggest that disease
    risk variants affect gene regulation: They are enriched in regions
    with regulatory function active in relevant cell populations;[16,17]
    and they appear to drive changes to gene expression.[18] This may
    help explain the modest amount of risk each variant encodes, as
    changes to gene regulation are likely to be milder in phenotypic
    effect than those altering the encoded protein's structure.

4. *Loci can harbor multiple independent effects.* In many loci, there
    appear to be multiple uncorrelated variants associated to dis-
    ease. By *conditional analysis* — regressing out the effect of the
    most associated marker — several independent statistical asso-
    ciations can be found in each region.[9,18] These observations are
    consistent with two explanations that have yet to be conclusively
    tested: Either multiple genes in a region are effected, or there
    are several, cumulative effects on a single gene. In one case at
    least we know it is the latter: The locus encoding the IRF5 gene
    harbors three independent associations to SLE; in combination,
    they produce an IRF5 transcript with an alternate first exon, a
    30-base insertion in exon 6 and an altered 3'UTR and poly-
    adenosine tail, which appears to mediate risk.[19] Further analy-
    ses, in combination with the regulatory findings described above
    and innovative experimental frameworks, will show how repre-
    sentative this finding is.

5. *There is substantial sharing of risk variants across diseases.* As more
    and more loci were associated to one or another disease, it
    became obvious that many of them were shared across multiple
    diseases.[20] When loci are grouped by the diseases to which they
    mediate risk, they appear to contain genes participating in the
    same pathways.[21] This suggests that perturbation of a pathway
    may result in elevated risk to more than one disease — and that
    targeting that pathway therapeutically may be useful in more
    than one disease, too.[22]

Beyond these general observations on the architecture of disease, several specific biological insights have already emerged from genetic studies of inflammatory and autoimmune diseases. We briefly review two examples here, which illustrate the potential for novel biological insights stemming from GWAS. In multiple sclerosis, the loci identified by GWAS studies overwhelmingly encode genes specific to subsets of CD4$^+$ helper T cells.[10,23] The pathogenic role of these cell types in the disease is supported by two complementary strands of evidence: First, MS risk SNPs are enriched in DNase I hypersensitivity sites active in helper T cell populations;[15] second, these SNPs drive changes to gene expression preferentially in these cells.[17] Thus, although neuro-pathological and physiological observations place CD8$^+$ T cells and B cells in the brain during overt, long-standing disease, it is becoming obvious that the disease is initiated in CD4$^+$ T cells — important information for development of early intervention therapeutics to block disease initiation, rather than the broad-spectrum immune suppression regimens currently in use.

In Crohn's disease (one of the two principal subtypes of IBD), early GWAS studies detected an association to a region on chromosome 2 encoding the gene ATG16L1.[24] In humans, the ATG16L1 protein has a critical role in completing the double-membrane envelope around bacterial cells present in the cytoplasm of epithelial cells and their subsequent trafficking to lysosomes for degradation. The IBD-associated variant induces a defect in the enveloping process, stranding engulfed bacteria in the cytoplasm and creating chronic pro-inflammatory conditions.[25] In combination with the previous discovery of genetic associations to the loci encoding the NOD2 and IRGM genes — also involved in bacterial autophagy — this finding has, for the first time, implicated this process as central to IBD pathogenesis.[26]

3. Genetic Mapping in Infectious Diseases

Infectious disease research has mostly focused on pathogen biology, leading to remarkable breakthroughs in treatment and eradication of bacterial, viral and fungal infections. More recently, GWAS has been used to identify host susceptibility genes across a spectrum of diseases (Table 1), including visceral leishmaniasis,[27] hepatitis B and C infection, progression,[28–30] viral clearance[31,32] and treatment response,[33,34] and HIV infection and virus titer control.[35] Notable difficulties in such studies include identifying exposed but uninfected individuals as suitable controls, and ascertaining samples in regions of the world where such infections are common but medical infrastructure is lacking.

Whilst early in the discovery phase of GWAS, intriguing insights are already emerging. Two illustrative examples are the discovery that variants in IL28, which encodes the type III interferon IFN-λ3, mediate both response to hepatitis C treatment and the spontaneous clearance of the virus in the absence of treatment.[31,32] In HIV, deletion of 32 base-pairs of the gene CCR5, which encodes the C–C chemokine receptor type 5 protein (also known as CD195), confers protection to infection of M-tropic strains of HIV-1. CCR5 is used by the virus to infect host T cells, which is impaired by the deletion variant (CCR5-Δ32). A haplotype spanning the class I human leukocyte antigen (HLA) genes in the major histocompatibility locus (MHC) on chromosome 6 is also associated with HIV viral control. This haplotype, *HLA-B*57-01*, encodes specific amino acid changes to the floor of the peptide-binding groove of *HLA-B*; it also confers sensitivity to the HIV drug abacavir and to drug-induced skin inflammatory disease, suggesting a broader functional role in immune system function.[35]

4. Genetic Mapping of Immune Function and Response Phenotypes

In addition to diseases relevant to the immune system, genetic determinants of various aspects of broader immune function are now being sought. Variants in the MHC region influence the ratio of CD4$^+$ to CD8$^+$ T cells at baseline, and have been suggested to confer susceptibility to a number of immune-related diseases including T1D and HIV infection.[36] Immune response also appears to be under partial genetic control: Response to antigen challenge through vaccination against several viruses,[37,38] most notably diverse strains of influenza,[39,40] have been mapped to the MHC region on chromosome 6 and likely have multiple other associations. These emerging studies promise to increase our understanding of immune function, as demonstrated by the examples above.[41]

5. Conclusions

Over the last 15 years, genetic mapping of diverse traits related to the human immune system have revolutionized our understanding of immune function. This has been catalyzed by the sequencing of the human genome, cataloging of common genetic variation across populations and development of high-throughput genotyping technologies. Whilst it is impossible in this space to provide an exhaustive discussion of all findings, we have already gained valuable insights. The genetic and phenotypic complexity of immune function of disease is more clear than ever. The broader MHC region on chromosome 6 has a pre-eminent role in many traits, as is clear from associations to almost every aspect of immune function, even though the biological basis is still unclear 40 years after the first reports. New and unexpected aspects of pathobiology have been revealed by genetic results and are likely but the first wave of such observations, requiring new models and modes of thought.

Most important, perhaps, is the overall observation of the sheer complexity of these traits. Thousands of variants affecting hundreds of genes mediate disease risk and immune function; these are neither necessary nor sufficient for phenotype, and in most cases appear to impact gene regulation rather than function. To test these factors at such a scale requires new experimental paradigms capable of studying the activity and interplay of many genes simultaneously, and across many relevant cell subpopulations. A coherent approach of systems-level immunology will be required to unravel how risk accumulates in relevant molecular pathways and to eventually identify modifiers of the underlying processes that may be targeted by therapies. This complex process must begin at understanding the biological effects of these alleles at baseline and in response to immune challenge.

The prospect may seem daunting, but the assault on understanding the immune system and its role in disease is now driven by a wealth of robust findings which require imaginative, but careful, interpretation.

Glossary

- Common disease, common variant (CDCV) hypothesis — The hypothesis that common disease risk is driven by a number of common variants, each of which explains a small fraction of the total heritability (and are thus neither necessary nor sufficient by themselves).
- Conditional analysis — regressing out the primary association in a locus to establish if there are additional, independent signals.
- *de novo* mutation — a mutation arising in an individual for the first time, usually during meiotic recombination one of the parents' sex cells.
- Genetic architecture — the number of genetic variants underlying a trait and the distribution of effect sizes.
- Genetic mapping — correlating genotype to phenotype in a population sample to identify regions of the genome driving phenotype.
- Genetic variant — a polymorphism observed in the general population. Generally common and thus found in unrelated individuals.
- Genome-wide association study — a method for genetic mapping in a population, where minor allele frequency is compared across affection status (binary traits) or correlated to phenotype value (continuous traits). The dominant study format in modern genetics due to the sample sizes achievable.
- Genotyping — establishing the alleles carried at each marker for an individual. Chip-based single-base extension assays (multiplexed in millions) are gradually being replaced by next-generation sequencing as the dominant form.
- Heritability — the proportion of variance in a phenotype attributable to genetic variants in a population.
- Linkage analysis — a method for genetic mapping in families, by comparing the segregation of each allele of a marker with a trait.
- Meiosis — cell division that leads to haploid daughter cells. The process by which sperm and ova are produced.

- Monogenic trait — a trait where a mutation of a single gene is sufficient to induce phenotype. Note that these traits can be *heterogeneous*: Mutations in different genes can cause the same phenotype.
- Mutation — a rare, usually deleterious genetic variant. Mutations are understood to be private to families or isolated populations.
- Polygenic trait — a complex trait where multiple genetic variants each explain a portion of the heritability.
- Statistical power — the chance (as a percentage) that a given study design will detect an effect if it is truly there. In genetic studies, a combination of sample size, marker frequency, effect size of the risk variant and level of significance desired.
- Trait or phenotype — a property of individuals in a population. Includes binary traits such as disease status and continuous measures such as height, blood pressure, and antibody titre.

References

1. Frazer KA, Murray SS, Schork NJ *et al.* (2009) Human genetic variation and its contribution to complex traits. *Nat Rev Genet* **10:** 241–251.
2. Gibson G. (2012) Rare and common variants: Twenty arguments. *Nat Rev Genetics* **13:** 135–145.
3. International HapMap Consortium (2005) A haplotype map of the human genome. *Nature* **437:** 1299–1320.
4. International HapMap Consortium (2010) Integrating common and rare genetic variation in diverse human populations. *Nature* **467:** 52–58.
5. Hirschhorn JN, Daly MJ. (2005) Genome-wide association studies for common diseases and complex traits. *Nat Rev Genet* **6:** 95–108.
6. Altshuler D, Daly MJ, Lander ES (2008) Genetic Mapping in Human Disease. *Science* **322:** 881–888.
7. Samarghitean C, Väliaho J, Vihinen M. (2007) IDR knowledge base for primary immunodeficiencies. *Immunome Res* **3:** 6.
8. Buckley RH. (2004) Molecular defects in human severe combined immunodeficiency and approaches to immune reconstitution. *Annu Rev Immunol.* **22:** 625–655.
9. Jostins L *et al.* (2012) Host–microbe interactions have shaped the genetic architecture of inflammatory bowel disease. *Nature* **491:** 119–124.
10. International Multiple Sclerosis Genetics Consortium (IMSGC) *et al.* (2013) Analysis of immune-related loci identifies 48 new susceptibility variants for multiple sclerosis. *Nat Genet* **45:** 1353–1360.

11. Tsoi LC *et al.* (2012) Identification of 15 new psoriasis susceptibility loci highlights the role of innate immunity. *Nat Genet* **44:** 1341–1348.

12. Genetic Analysis of Psoriasis Consortium & the Wellcome Trust Case Control Consortium 2 *et al.* (2010) A genome-wide association study identifies new psoriasis susceptibility loci and an interaction between HLA-C and ERAP1. *Nat Genet* **42:** 985–990.

13. Okada Y *et al.* (2013) Genetics of rheumatoid arthritis contributes to biology and drug discovery. *Nature* **506:** 376–381.

14. Cunninghame Graham DS *et al.* (2011) Association of NCF2, IKZF1, IRF8, IFIH1, and TYK2 with Systemic Lupus Erythematosus. *PLoS Genet* **7:** e1002341.

15. Barrett JC *et al.* (2009) Genome-wide association study and meta-analysis find that over 40 loci affect risk of type 1 diabetes. *Nat Genet* **41:** 703–707.

16. Maurano MT *et al.* (2012) Systematic localization of common disease-associated variation in regulatory DNA. *Science* **337:** 1190–1195.

17. Trynka G *et al.* (2013) Chromatin marks identify critical cell types for fine mapping complex trait variants. *Nat Genet* **45:** 124–130.

18. Raj T *et al.* (2014) Polarization of the effects of autoimmune and neurodegenerative risk alleles in leukocytes. *Science* **344:** 519–523.

19. Patsopoulos NA, the Bayer Pharma MS Genetics Working Group, the Steering Committees of Studies Evaluating IFNβ-1b and a CCR1-Antagonist, ANZgene Consortium, GeneMSA, International Multiple Sclerosis Genetics Consortium & de Bakker, P. I. W. (2011) Genome-wide meta-analysis identifies novel multiple sclerosis susceptibility loci. *Ann Neurol.* **70:** 897–912.

20. Graham RR *et al.* (2007) Three functional variants of IFN regulatory factor 5 (IRF5) define risk and protective haplotypes for human lupus. *Proc Natl Acad Sci USA* **104:** 6758–6763.

21. Becker KG *et al.* (1998) Clustering of non-major histocompatibility complex susceptibility candidate loci in human autoimmune diseases. *Proc Natl Acad Sci USA* **95:** 9979–9984.

22. Cotsapas C *et al.* (2011) Pervasive sharing of genetic effects in autoimmune disease. *PLoS Genet* **7:** e1002254.

23. Voight BF, Cotsapas C. (2012) Human genetics offers an emerging picture of common pathways and mechanisms in autoimmunity. *Curr Opin Immunol* **24:** 552–557.

24. International Multiple Sclerosis Genetics Consortium *et al.* (2011) Genetic risk and a primary role for cell-mediated immune mechanisms in multiple sclerosis. *Nature* **476:** 214–219.

25. Barrett JC. *et al.* (2008) Genome-wide association defines more than 30 distinct susceptibility loci for Crohn's disease. *Nat Genet* **40**: 955–962.

26. Kuballa P, Huett A, Rioux JD *et al.* (2008) Impaired autophagy of an intracellular pathogen induced by a Crohn's disease associated ATG16L1 variant. *PLoS ONE* **3**: e3391.

27. Xavier RJ, Rioux JD. (2008) Genome-wide association studies: A new window into immune-mediated diseases. *Nat Rev Immunol* **8**: 631–643.

28. LeishGEN Consortium *et al.* (2013) Common variants in the HLA-DRB1-HLA-DQA1 HLA class II region are associated with susceptibility to visceral leishmaniasis. *Nat Genet* **45**: 208–213.

29. Wang D, Yu H, Zhang P *et al.* (2013) Genetic variants in STAT4 and HLA-DQ genes confer risk of Hepatitis B virus-related hepatocellular carcinoma. *Nat Genet* **45**: 72–75 (2013).

30. Zhang H *et al.* (2010) Genome-wide association study identifies 1p36. 22 as a new susceptibility locus for hepatocellular carcinoma in chronic Hepatitis B virus carriers. *Nat Genet* **42**(9): 755–758.

31. Yang J *et al.* (2012) GWAS identifies novel susceptibility loci on 6p21. 32 and 21q21. 3 for hepatocellular carcinoma in chronic Hepatitis B virus carriers. *PLoS Genet.*

32. Muir AJ, Sulkowski M, McHutchison JG *et al.* (2009) Genetic variation in IL28B predicts Hepatitis C treatment-induced viral clearance. *Nature* **461**, 399–401 (17 September 2009).

33. Tobler LH, Busch MP, McHutchison JG *et al.* (2009) Genetic variation in IL28B and spontaneous clearance of Hepatitis C virus. *Nature* 2009 Oct 8; **461**(7265): 798–801.

34. Fellay J *et al.* (2010) ITPA gene variants protect against anaemia in patients treated for chronic Hepatitis C. *Nature* **464**, 405–408 (18 March 2010).

35. Suppiah V *et al.* (2009) IL28B is associated with response to chronic Hepatitis C interferon-α and ribavirin therapy. *Nat Genet* 1–6.

36. The International HIV Controllers Study. (2010) The major genetic determinants of HIV-1 control affect HLA Class I peptide presentation. *Science* **330**: 1551–1557.

37. Ferreira MAR *et al.* (2010) Quantitative trait loci for CD4: CD8 lymphocyte ratio are associated with risk of type 1 diabetes and HIV-1 immune control. *Am J Hum Genet* **86**: 88–92.

38. Haralambieva IH *et al.* (2011) Common SNPs/haplotypes in IL18R1 and IL18 genes are associated with variations in humoral immunity to smallpox vaccination in Caucasians and African Americans. *J Infect Dis* **204**: 433–441.

39. Ovsyannikova IG *et al.* (2012) Genome-wide association study of antibody response to smallpox vaccine. *Vaccine* **30:** 4182–4189.
40. Majumder PP *et al.* (2013) Genomic correlates of variability in immune response to an oral cholera vaccine. *Eur J Hum Genet* **21:** 1000–1006.
41. Pajewski NM *et al.* (2012) A genome-wide association study of host genetic determinants of the antibody response to anthrax vaccine adsorbed. *Vaccine* **30:** 4778–4784.
42. Poland GA, Ovsyannikova IG, Jacobson RM. (2008) Immunogenetics of seasonal influenza vaccine response. *Vaccine* **26:** D35–D40.
43. Yang S-K *et al.* (2014) Genome-wide association study of Crohn's disease in Koreans revealed three new susceptibility loci and common attributes of genetic susceptibility across ethnic populations. *Gut.* 2014 Jan; **63**(1): 80–7.
44. Gateva V *et al.* (2009) A large-scale replication study identifies TNIP1, PRDM1, JAZF1, UHRF1BP1 and IL10 as risk loci for systemic lupus erythematosus. *Nat Genet* **41:** 1228–1233.
45. Yang W, Tang H, Zhang Y *et al.* (2013) Meta-analysis Followed by Replication Identifies Loci in or near CDKN1B, TET3, CD80, DRAM1, and ARID5B as Associated with Systemic Lupus Erythematosus in Asians. *Am J Hum Genet* **92:** 41–51.
46. Okada Y *et al.* (2012) Meta-analysis identifies nine new loci associated with rheumatoid arthritis in the Japanese population. *Nat Genet* **44:** 511–516.
47. Trynka G *et al.* (2011) Dense genotyping identifies and localizes multiple common and rare variant association signals in celiac disease. *Nat Genet* **43:** 1193–1201.
48. Paternoster L *et al.* (2012) Meta-analysis of genome-wide association studies identifies three new risk loci for atopic dermatitis. *Nat Genet* **44:** 187–192.
49. Hirota T *et al.* (2012) Genome-wide association study identifies eight new susceptibility loci for atopic dermatitis in the Japanese population. *Nat Genet* **44:** 1222–1226.
50. Hinds DA *et al.* (2013) A genome-wide association meta-analysis of self-reported allergy identifies shared and allergy-specific susceptibility loci. *Nat Genet* **45:** 907–911.
51. Bønnelykke K *et al.* (2013) Meta-analysis of genome-wide association studies identifies ten loci influencing allergic sensitization. *Nat Genet* **45:** 902–906.

52. Ferreira RC *et al.* (2012) High-density SNP mapping of the HLA region identifies multiple independent susceptibility loci associated with selective IgA deficiency. *PLoS Genet* **8:** e1002476.
53. Khor CC *et al.* (2011) Genome-wide association study identifies susceptibility loci for dengue shock syndrome at MICB and PLCE1. *Nat Genet* **43:** 1139–1141.
54. Timmann C *et al.* (2012) Genome-wide association study indicates two novel resistance loci for severe malaria. *Nat* **489:** 443–446.
55. Thye T *et al.* (2012) Common variants at 11p13 are associated with susceptibility to tuberculosis. *Nat Genet* **44:** 257–259.
56. Thye T *et al.* (2010) Genome-wide association analyses identifies a susceptibility locus for tuberculosis on chromosome 18q11.2. *Nat Genet* **42:** 739–741.
57. Zhang F *et al.* (2011) Identification of two new loci at IL23R and RAB32 that influence susceptibility to leprosy. *Nat Genet* **43:** 1247–1251.

Index